Longman Mathematical Texts

Random variables

Longman Mathematical Texts

Random variables L.E. Clarke

Electricity C.A. Coulson

Continuum mechanics A.J.M. Spencer

Elasticity R.J. Atkin and N. Fox

Set theory and abstract algebra T.S. Blyth

Elementary mathematical analysis I.T. Adamson

Electricity C.A. Coulson and T.J.M. Boyd

Optimization A.H. England and W.A. Green

Functions of a complex variable E.G. Phillips

Edited by **Alan Jeffrey** and **Iain Adamson**

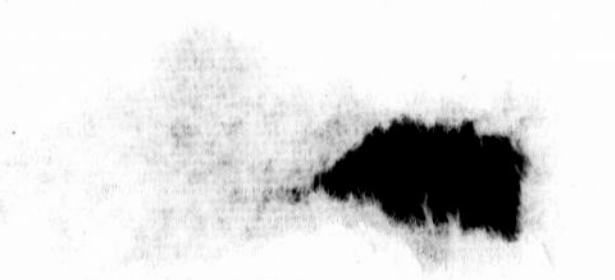

Longman Mathematical Texts

Random variables

L.E. Clarke

Senior Lecturer in Mathematics,
University of East Anglia

Longman
London and New York

Longman Group Limited

London

Associated companies, branches and representatives throughout the world

Published in the United States of America/by Longman Inc., New York

First published 1975

ISBN 0 582 44277 X

Library of Congress Catalog Card Number 73-85678

Printed in Hungary

Preface

The aim of this book is to provide an introduction to advanced probability theory suitable for third-year university students of mathematics. It deals in a rigorous manner with such topics as probability space, random variable, distribution, expected value, independence and characteristic function, and ends with proofs of the three principal limit theorems for independent and identically distributed random variables – the weak and strong laws of large numbers, and the central limit theorem.

The reader is assumed to be reasonably familiar with the analysis normally taught in the first half of an honours course in mathematics. This does not usually include Lebesgue theory, and so an appendix on it has been provided which, it is hoped, will make the one or two references to it meaningful.

Strictly speaking, no previous acquaintance with probability theory is needed. However, since some knowledge of the experimental background to probability theory and its basic intuitive ideas provides motivation for, and signposts along, the road we shall follow, any understanding, however slight, which the reader may already have of such notions as "event", "probability", "random variable" and "independence" will serve him well when they are treated in a more abstract and mathematical way. Such understanding may easily be gained from one of the many texts available on elementary probability theory.

The exercises are important, not only as a means of developing a new skill, but also because they contain many important results, some of which will be needed later in the text.

Theorem x of chapter y will normally be referred to as theorem x in chapter y itself, and as theorem $y{\cdot}x$ elsewhere, and similarly for theorems in the appendices and for exercises.

Finally, I should like to express my thanks to the University of East Anglia for a grant of study leave which helped me to complete the writing of the book, and to Dr C. G. C. Pitts for his critical reading of the text and his helpful comments. As a result of the latter many deficiencies have been removed from an earlier draft. For such as remain, I alone am responsible, and should be grateful if my attention were to be drawn to them.

L. E. Clarke

Contents

Preliminaries

(1) The set of all real numbers will be denoted by R (*the real line*).

The reader is reminded that the real numbers are finite by definition.

(2) If $a, b \in R$ and $a < b$, the closed and open intervals in R with end-points a and b will be denoted by $[a, b]$ and (a, b) respectively, i.e.

$$[a, b] = \{x: a \leqslant x \leqslant b\}$$

and

$$(a, b) = \{x: a < x < b\}.$$

This notation will be extended in the obvious manner to intervals such as

$$(a, b] = \{x: a < x \leqslant b\}$$

and

$$(-\infty, b] = \{x: -\infty < x \leqslant b\} = \{x: x \leqslant b\}.$$

(3) If $x_1, x_2, \ldots$ is a sequence of real numbers, either of the statements

$$x_n \to x \quad \text{as} \quad n \to \infty, \qquad \lim_{x_n \to \infty} x_n = x$$

will imply that x itself is a real number, and therefore finite.

If infinite limits are allowed, this will be explicitly indicated by some such wording as

$$x_n \to x \quad \text{as} \quad n \to \infty, \quad \text{where } x \text{ may be infinite.}$$

For example, if $c \geqslant 0$, then $c^n \to f(c)$ as $n \to \infty$, where $f(c)$ may be $+\infty$.

(4) If c is a positive real number, $\sqrt{c}$ will always denote the positive root of $x^2 = c$.

(5) A set will be said to be *enumerable* if it can be put into one-one correspondence with the set of positive integers, and *countable* if it is finite or enumerable.

1
Sets

This first chapter will contain the principal definitions and results relating to sets which will be of use later in this book.

The union and the intersection of two sets A and B will be denoted by $A \cup B$ and $A \cap B$ respectively, and this notation will be extended in the obvious manner to the union and the intersection of the sets A_t $(t \in T)$, where T is a non-empty index set (index sets will always be assumed to be non-empty). Thus the union of the sets A_t $(t \in T)$ will be denoted by $\bigcup_{t \in T} A_t$ (or $\bigcup_t A_t$, or even $\bigcup A_t$, if there is no possibility of confusion), and similarly for the intersection. More precisely

$$\bigcup_{t \in T} A_t = \{\omega : \omega \in A_t \text{ for at least one } t \in T\}$$

(i.e. the set of those elements ω such that ω belongs to A_t for at least one t belonging to T), and

$$\bigcap_{t \in T} A_t = \{\omega : \omega \in A_t \text{ for every } t \in T\}.$$

For any two sets A and B the set of those elements which belong to A but not to B will be denoted by $A - B$, i.e.

$$A - B = \{\omega : \omega \in A \text{ and } \omega \notin B\}.$$

In particular, if all the sets under consideration are regarded as subsets of some *space* Ω (a fixed non-empty "universal" set) then the complement of A is defined to be $\Omega - A$, and will be denoted by A^c. Thus $A - B = A \cap B^c$.

The empty set will be denoted by $\emptyset$.

The reader is assumed to be familiar with the basic formulae of the algebra of sets, such as

$$A \cap (B \cup C) = (A \cap B) \cup (A \cap C)$$

and

$$\left(\bigcup_t A_t\right)^c = \bigcap_t A_t^c.$$

Usually sets will be denoted by Italic capitals $A, B, C, \ldots$, with or without affixes. (The converse of this statement is false; for example, P will denote a probability measure (see Chapter 2), and X, Y, Z will denote random variables (see Chapter 4).) In particular, the set of all real numbers ("the real line") will be denoted by R. Exceptionally, a general space will be denoted by Ω. *Classes* of sets will be denoted by script capitals $\mathscr{A}$, $\mathscr{B}$, $\mathscr{C}, \ldots$. The class of all subsets of a space Ω will be denoted by $\mathscr{P}(\Omega)$.

Two particular types of class will play a large part in what follows. Firstly, in Chapter 2 we shall meet the class of events corresponding to the performance of some experiment. This can be regarded as a class of sets in a space Ω, and the discussion in that chapter will show that it is reasonable to postulate that the class is closed with respect to the operations of taking the complement of any one of its sets, and forming the union of any sequence of its sets. We are thus led to the definition of a sigma-algebra given below. Secondly, in Chapter 4 we shall find that intervals of the form $(a, b]$ occur in a very natural manner in connection with the distribution functions there introduced. Such intervals can also be regarded as the basic building bricks of the real line, since R can be expressed as a union of disjoint intervals of this form, for example

$$R = \bigcup_{n=-\infty}^{\infty} (n, n+1].$$

Note that the union of the corresponding open intervals $(n, n+1)$ would omit the integers, and that of the corresponding closed intervals $[n, n+1]$ would cover them twice over. Now if A and B are two intervals of the form $(a, b]$ then $A \cap B$, if not empty, is an interval of this form, and $A-B$, if not empty, is the union of at most two intervals of this form. These last two properties, with a slight modification of the second, suggest the formal definition of a semi-algebra.

Definitions. Let $\mathscr{C}$ be a class of sets in a space Ω.

(i) $\mathscr{C}$ is a *sigma-algebra* (σ-algebra) in Ω if

(a) $\Omega \in \mathscr{C}$;

(b) whenever $A \in \mathscr{C}$, then also $A^c \in \mathscr{C}$;

(c) whenever $A_1, A_2, \ldots \in \mathscr{C}$, then also $\bigcup_1^{\infty} A_n \in \mathscr{C}$.

(ii) $\mathscr{C}$ is a *seml-algebra* in Ω if

(a) $\Omega \in \mathscr{C}$;
(b) whenever $A \in \mathscr{C}$, then A^c can be expressed as the union of a *finite* number of disjoint sets of $\mathscr{C}$;
(c) whenever $A, B \in \mathscr{C}$, then also $A \cap B \in \mathscr{C}$.

The terminology is by no means fixed. In particular, "sigma-field" and "Borel field" are both used in place of "sigma-algebra". However, the term "sigma-algebra" is the one now most generally preferred.

The reader should prove the following results (if he has difficulty in doing so, he may refer to the proof of Lemma 2.1):

(i) if $\mathscr{C}$ is a sigma-algebra then

(a) whenever $A_1, A_2, \ldots \in \mathscr{C}$, then also $\bigcap_1^\infty A_n \in \mathscr{C}$;
(b) $\emptyset \in \mathscr{C}$;
(c) whenever $A, B \in \mathscr{C}$, then also $A - B \in \mathscr{C}$;
(d) $\mathscr{C}$ is a semi-algebra.

(ii) if $\mathscr{C}$ is a semi-algebra then $\emptyset \in \mathscr{C}$.

Trivial examples of sigma-algebras in a space Ω are $\{\emptyset, \Omega\}$ and $\mathscr{P}(\Omega)$. In an obvious sense of the words, they are respectively the smallest and the largest sigma-algebra in Ω.

A particularly important semi-algebra, which we shall meet again later, occurs when $\Omega = R$. Let $\mathscr{I}$ be the class of all sets

$$\begin{aligned} (a, b] & \quad (-\infty < a < b < \infty), \\ (-\infty, b] & \quad (-\infty < b < \infty), \\ (a, \infty) & \quad (-\infty < a < \infty), \end{aligned}$$

together with $\emptyset$ and R. Then, as the reader will easily verify, $\mathscr{I}$ is a semi-algebra of sets in R.

We shall now prove three theorems which will be of frequent use.

Theorem 1 *Let $\mathscr{C}$ be any non-empty class of sets in a space Ω. Then there exists one and only one sigma-algebra $\mathscr{F}$ in Ω such that*

(i) *$\mathscr{F} \supseteq \mathscr{C}$ (i.e. $\mathscr{F}$ contains all the sets of $\mathscr{C}$);*
(ii) *if $\mathscr{F}_1$ is any sigma-algebra in Ω for which $\mathscr{F}_1 \supseteq \mathscr{C}$ then $\mathscr{F}_1 \supseteq \mathscr{F}$.*

Note. $\mathscr{F}$ is the "smallest" sigma-algebra in Ω which contains all the sets of $\mathscr{C}$. It is referred to as the sigma-algebra *generated* by $\mathscr{C}$, and is denoted by $\sigma(\mathscr{C})$.

Proof. We show first that there exists at least one sigma-algebra $\mathscr{F}$ having the properties (i) and (ii).

There do exist sigma-algebras in Ω which contain all the sets of $\mathscr{C}$ (for example, $\mathscr{P}(\Omega)$). Let us define $\mathscr{F}$ to be the class of those sets which belong to all such sigma-algebras; thus

$$\mathscr{F} = \bigcap \mathscr{S},$$

where $\mathscr{S}$ runs through all those sigma-algebras in Ω which contain all the sets of $\mathscr{C}$.

Clearly $\mathscr{F}$ satisfies (i) and (ii). Also $\mathscr{F}$ is a sigma-algebra. For firstly $\Omega \in \mathscr{F}$ (since $\Omega \in \mathscr{S}$ for all $\mathscr{S}$). Next suppose that $A \in \mathscr{F}$. Then, for all $\mathscr{S}$, $A \in \mathscr{S}$ (by definition of $\mathscr{F}$), and so $A^c \in \mathscr{S}$; therefore $A^c \in \mathscr{F}$ (again by definition of $\mathscr{F}$). Finally, if $A_1, A_2, \ldots \in \mathscr{F}$ it can be shown similarly that $\bigcup_1^\infty A_n \in \mathscr{F}$. Thus $\mathscr{F}$ is a sigma-algebra.

We now show that $\mathscr{F}$ is unique.

Suppose that $\mathscr{F}$, $\mathscr{F}^*$ are sigma-algebras in Ω such that $\mathscr{F}$ satisfies (i), (ii) above and $\mathscr{F}^*$ satisfies

(i*) $\mathscr{F}^* \supseteq \mathscr{C}$;
(ii*) if $\mathscr{F}_1$ is any sigma-algebra in Ω for which $\mathscr{F}_1 \supseteq \mathscr{C}$ then $\mathscr{F}_1 \supseteq \mathscr{F}^*$.

Then $\mathscr{F}^* \supseteq \mathscr{C}$ (by (i*)), and so $\mathscr{F}^* \supseteq \mathscr{F}$ (by (ii)). Similarly $\mathscr{F} \supseteq \mathscr{F}^*$, and so $\mathscr{F} = \mathscr{F}^*$.

Note. In Theorem 1 it was not strictly necessary to assume that $\mathscr{C}$ was non-empty. If, in fact, $\mathscr{C}$ is empty, then the generated sigma-algebra is easily seen to be $\{\emptyset, \Omega\}$. However, many readers will regard non-empty classes of sets as being somewhat less nebulous, and in what follows Theorem 1 will be applied only to such classes.

An important generated sigma-algebra occurs when $\Omega = R$ and $\mathscr{C}$ is the class of open sets in R. Then the corresponding generated sigma-algebra $\sigma(\mathscr{C})$ is the class $\mathscr{B}$ of *Borel sets* in R. Roughly speaking, $\mathscr{B}$ consists of all those sets in R which may be obtained from the intervals by repeatedly forming complements and countable unions. It contains all the sets which occur naturally in analysis.

Theorem 2 *Let $\mathscr{I}$ be the class of sets*

$$(-\infty, x] \quad (-\infty < x < \infty)$$

in R. Then $\mathscr{B} = \sigma(\mathscr{I})$.

Proof. Let $\mathcal{G}$ be the class of open sets in R. Then for all real x

$$(x, \infty) \in \mathcal{G}$$

and so, since $\mathcal{B} = \sigma(\mathcal{G}) \supseteq \mathcal{G}$ (by Theorem 1(i)),

$$(x, \infty) \in \mathcal{B}$$

Therefore $$(-\infty, x] = R - (x, \infty) \in \mathcal{B}$$

(because $\mathcal{B}$ is a sigma-algebra). Therefore $\mathcal{J} \subseteq \mathcal{B}$, and so (by Theorem 1(ii)),

$$\sigma(\mathcal{J}) \subseteq \mathcal{B}$$

(using again the fact that $\mathcal{B}$ is a sigma-algebra).

To obtain the reverse inclusion $\mathcal{B} \subseteq \sigma(\mathcal{J})$ it suffices to prove that $\mathcal{G} \subseteq \sigma(\mathcal{J})$, from which it follows that $\mathcal{B} = \sigma(\mathcal{G}) \subseteq \sigma(\mathcal{J})$ (by Theorem 1(ii)).

Let G be any non-empty open set in R, and let x be any point of G. Then there exists a number $\delta > 0$ such that

$$(x-\delta, x+\delta) \subseteq G$$

(because G is open), and then there exist rational numbers r, s such that

$$x-\delta < r < x < s < x+\delta.$$

Thus to each $x \in G$ correspond rational numbers r, s such that

$$x \in (r, s] \subseteq G.$$

Now since the set of rational numbers is enumerable, so is the set of all ordered pairs (r, s) of rational numbers. It follows that the subset of those pairs (r, s) for which $r < s$ is also enumerable. Therefore the set of all those intervals $(r, s]$ for which r and s are rational is enumerable, and so the subset consisting of those intervals contained in G is also enumerable (it is clearly not finite). Let it consist of the intervals

$$(r_n, s_n] \qquad (n = 1, 2, \ldots).$$

Since, as we have already shown, any point of G is contained in at least one of these intervals, and since these intervals are all contained in G,

$$G = \bigcup_{n=1}^{\infty} (r_n, s_n].$$

Now, for $n = 1, 2, \ldots$

$$(r_n, s_n] = (-\infty, s_n] - (-\infty, r_n] \in \sigma(\mathcal{J}),$$

because $(-\infty, r_n], (-\infty, s_n] \in \mathscr{J} \subseteq \sigma(\mathscr{J})$, and $\sigma(\mathscr{J})$ is a sigma-algebra. Therefore, by part (c) of the definition of a sigma-algebra,

$$G \in \sigma(\mathscr{J}).$$

The reader should note that it was essential to express G as the union of an enumerable class of sets of $\sigma(\mathscr{J})$ for this last step to be justified. It would not have sufficed to express G as the union of those intervals $(r, s]$ for which $(r, s] \subseteq G$ and r and s are real, for the class of such intervals is not enumerable.

Since also $\emptyset \in \sigma(\mathscr{J})$, it follows that $G \in \sigma(\mathscr{J})$ for every open set G in R, and so $\mathscr{G} \subseteq \sigma(\mathscr{J})$; this completes the proof.

The notation $\mathscr{I}$, $\mathscr{J}$, $\mathscr{B}$ introduced above will be used throughout this book. The corresponding classes in n-dimensional Euclidean space R^n will be denoted by $\mathscr{I}^n$, $\mathscr{J}^n$, $\mathscr{B}^n$; they are defined by

$$\mathscr{I}^n = \{I_1 \times I_2 \times \ldots \times I_n : I_1, I_2, \ldots, I_n \in \mathscr{I}\},$$
$$\mathscr{J}^n = \{J_1 \times J_2 \times \ldots \times J_n : J_1, J_2, \ldots, J_n \in \mathscr{J}\},$$
$$\mathscr{B}^n = \text{the sigma-algebra generated by the class of open sets in } R^n.$$

The sets of $\mathscr{B}^n$ are the *Borel sets* in R^n.

The reader should note that, although R^n is the set of all ordered n-tuples $(x_1, x_2, \ldots, x_n)$ of elements of R, $\mathscr{B}^n$ is not the set of all ordered n-tuples $(B_1, B_2, \ldots, B_n)$ of elements of $\mathscr{B}$. Thus the notation $\mathscr{B}^n$ is not an entirely happy one. However, it is now almost standard, and is universally understood. The notation $\mathscr{I}^n$, $\mathscr{J}^n$, again not entirely a happy one, is adopted for reasons of consistency. It can be shown that $\mathscr{I}^n$ is a semi-algebra and that $\mathscr{B}^n = \sigma(\mathscr{J}^n)$ (see Exercises 3 and 7).

The inverse image of a set

Let Ω, Ω' be two spaces, and suppose that

$$f : \Omega \to \Omega'$$

(i.e. to every $\omega \in \Omega$ corresponds a unique element $f(\omega)$ of Ω'). For any set $A' \subseteq \Omega'$ the *inverse image* of A' under f is defined to be

$$\{\omega : \omega \in \Omega \text{ and } f(\omega) \in A'\},$$

and is denoted by $f^{-1}(A')$ (note that $f^{-1}(A')$ is a set in Ω). Thus

$$\omega \in f^{-1}(A') \quad \text{if and only if} \quad f(\omega) \in A'.$$

In other words to each function $f:\Omega \to \Omega'$ corresponds a function f^{-1}: $\mathscr{P}(\Omega') \to \mathscr{P}(\Omega)$; the existence of this function f^{-1} in no way presupposes or implies that f has an inverse function in the ordinary sense (i.e. one from Ω' to Ω).

More generally, if $\mathscr{C}'$ is any non-empty class of sets in Ω', $f^{-1}(\mathscr{C}')$ is the class of inverse images under f of the sets of $\mathscr{C}'$, i.e.

$$f^{-1}(\mathscr{C}') = \{f^{-1}(A') : A' \in \mathscr{C}'\}.$$

It is a class of sets in Ω.

Theorem 3 *Let Ω, Ω' be two spaces, and suppose that*

$$f:\Omega \to \Omega'.$$

Then

(i) $f^{-1}(\Omega') = \Omega$ *and* $f^{-1}(\emptyset) = \emptyset$;
(ii) *if* $A' \subseteq \Omega'$,

$$f^{-1}(\Omega' - A') = \Omega - f^{-1}(A');$$

(iii) *if* $A'_t \subseteq \Omega'$ *for all* t *of some index set* T,

$$f^{-1}\left(\bigcup_{t \in T} A'_t\right) = \bigcup_{t \in T} f^{-1}(A'_t)$$

and

$$f^{-1}\left(\bigcap_{t \in T} A'_t\right) = \bigcap_{t \in T} f^{-1}(A'_t);$$

(iv) *if* $A', B' \subseteq \Omega'$,

$$f^{-1}(A' - B') = f^{-1}(A') - f^{-1}(B');$$

(v) *if* $A' \subseteq B' \subseteq \Omega'$,

$$f^{-1}(A') \subseteq f^{-1}(B');$$

(vi) *if* $\mathscr{B}'$ *and* $\mathscr{C}'$ *are two non-empty classes of sets in* Ω' *for which* $\mathscr{B}' \subseteq \mathscr{C}'$,

$$f^{-1}(\mathscr{B}') \subseteq f^{-1}(\mathscr{C}');$$

(vii) *if* $\mathscr{F}'$ *is a sigma-algebra in* Ω', $f^{-1}(\mathscr{F}')$ *is a sigma-algebra in* Ω;
(viii) *if* $\mathscr{F}$ *is a sigma-algebra in* Ω,

$$\{A' : A' \subseteq \Omega' \text{ and } f^{-1}(A') \in \mathscr{F}\}$$

(i.e. the class of those sets A' *in* Ω' *for which* $f^{-1}(A') \in \mathscr{F}$*) is a sigma-algebra in* Ω';

(ix) *if* $\mathscr{C}'$ *is a non-empty class of sets in* Ω',

$$\sigma\{f^{-1}(\mathscr{C}')\} = f^{-1}\{\sigma(\mathscr{C}')\}.$$

(In (ix) note that $\sigma(\mathscr{C}')$ is the sigma-algebra in Ω' generated by the class $\mathscr{C}'$, while $\sigma\{f^{-1}(\mathscr{C}')\}$ is the sigma-algebra in Ω generated by the class $f^{-1}(\mathscr{C}')$.)

Proof (i)–(viii) The proofs of (i)–(viii) follow at once from the appropriate definitions. As an example there follows the proof of (ii), the others being left to the reader.

$$\omega \in f^{-1}(\Omega' - A')$$

if and only if $\quad f(\omega) \in \Omega' - A',$

i.e. if and only if $\quad f(\omega) \notin A',$

i.e. if and only if $\quad \omega \notin f^{-1}(A'),$

i.e. if and only if $\quad \omega \in \Omega - f^{-1}(A').$

Thus

$$f^{-1}(\Omega' - A') = \Omega - f^{-1}(A').$$

(ix) The proof of this part lies a little deeper.

By Theorem 1, $\mathscr{C}' \subseteq \sigma(\mathscr{C}')$, and so, by (vi), $f^{-1}(\mathscr{C}') \subseteq f^{-1}\{\sigma(\mathscr{C}')\}$. But, by (vii), $f^{-1}\{\sigma(\mathscr{C}')\}$ is a sigma-algebra in Ω, and so, again by Theorem 1,

$$\sigma\{f^{-1}(\mathscr{C}')\} \subseteq f^{-1}\{\sigma(\mathscr{C}')\}. \tag{1}$$

Now let $\mathscr{S}' = \{A' : A' \subseteq \Omega' \text{ and } f^{-1}(A') \in \mathscr{F}\}$, where $\mathscr{F} = \sigma\{f^{-1}(\mathscr{C}')\}$. Then $\mathscr{C}' \subseteq \mathscr{S}'$ and by (viii), $\mathscr{S}'$ is a sigma-algebra in Ω'. Therefore, by Theorem 1, $\sigma(\mathscr{C}') \subseteq \mathscr{S}'$,

and so $\quad f^{-1}\{\sigma(\mathscr{C}')\} \subseteq f^{-1}(\mathscr{S}') \quad$ (by (vi))

$\quad\quad \subseteq \mathscr{F} \quad$ (by the definition of $\mathscr{S}'$).

Thus

$$f^{-1}\{\sigma(\mathscr{C}')\} \subseteq \sigma\{f^{-1}(\mathscr{C}')\}. \tag{2}$$

The result follows from (1) and (2).

Indicator functions

We conclude this chapter by defining the indicator function of a set.

Let A be any set in Ω. For any point ω of Ω let $I_A(\omega)$ be defined to be 1 or 0 according as $\omega \in A$ or $\omega \notin A$. Then the function

$$I_A : \Omega \to \{0, 1\}$$

is the *indicator function* of the set A. In particular, $I_\emptyset$ is the function which is identically 0 on Ω, and I_Ω is the function which is identically 1 on Ω.

There is clearly a one-one correspondence between sets in Ω and functions defined on Ω and taking their values in $\{0, 1\}$.

The term characteristic function is often used instead of indicator function, but it will not be so used in this book as it is reserved for a different purpose (see Chapter 8).

Exercises

1. Suppose that Ω is not countable. Show that the class

$$\{A\text{: one of } A, A^c \text{ is countable}\}$$

is a sigma-algebra in Ω.

2. Let $\mathscr{F}$ be a sigma-algebra in Ω, and let $A_1, A_2, \ldots$ be sets of $\mathscr{F}$. For $k = 1, 2, \ldots$ let

$$B_k = \{\omega : \omega \in A_n \text{ for at least } k \text{ values of } n\}.$$

(a) Show that

(i) $B_k \in \mathscr{F} \qquad (k = 1, 2, \ldots)$;
(ii) if $B = \{\omega : \omega \in A_n \text{ for an infinity of values of } n\}$, then

$$B = \bigcap_{k=1}^{\infty} B_k.$$

(b) Show also that

$$B = \bigcap_{k=1}^{\infty} (A_k \cup A_{k+1} \cup \ldots).$$

Hint. Note that $\omega \in A_n$ for an infinity of values of n if and only if

$$\omega \in A_k \cup A_{k+1} \cup \ldots$$

for every k.

(c) Deduce that $B \in \mathscr{F}$.

3. Prove that $\mathscr{I}^n$ is a semi-algebra in R^n.

Note. This exercise is typical of many occurring later in this book in that it generalises to R^n, where $n > 1$, a result holding in R. No new ideas are involved, and the reader should not encounter any difficulty (except, perhaps, with the notation). If he does, he should consider first the case $n = 2$ and draw a diagram.

4. If $A \neq \emptyset, \Omega$, prove that the sigma-algebra generated by $\{A\}$ is $\{\emptyset, A, A^c, \Omega\}$.

5. What is the sigma-algebra generated by the class $\{\{1, 2\}, \{1, 3\}\}$ if

(i) $\Omega = \{1, 2, 3, 4\}$,
(ii) $\Omega = \{1, 2, 3, 4, 5\}$?

6. Prove that the sigma-algebra generated by the class of closed sets in R^n is $\mathscr{B}^n$.

7. Prove that $\mathscr{B}^n = \sigma(\mathscr{J}^n)$.

Hint. Show first that any closed half-space, i.e. set typified by

$$\{(x_1, x_2, \ldots, x_n) : x_1 \leqslant c\},$$

is a countable union of sets of $\mathscr{J}^n$, and so $\sigma(\mathscr{J}^n)$ contains all closed half-spaces.

8. Show that the following subsets of R are Borel sets:

(i) The set of all rational numbers.

(ii) Any set of the form $\bigcap_{n=1}^{\infty} G_n$, where G_1 $G_2, \ldots$ are open.

(iii) Any set of the form $\bigcup_{n=1}^{\infty} F_n$, where $F_1, F_2, \ldots$ are closed.

Notes. (a) (i) is a special case of (iii).

(b) It is not easy to construct a set which is *not* a Borel set (such a set does not occur "naturally"). The reader requiring one should refer to any of the standard texts on Lebesgue theory for an example of a set which is not Lebesgue measurable, and therefore not a Borel set (see Appendix A).

9. Let $\mathscr{C}$, $\mathscr{C}^*$ and $\mathscr{C}_t$ $(t \in T)$ denote non-empty classes of sets in Ω. Prove that

(i) if $\mathscr{C} \subseteq \mathscr{C}^*$, then $\sigma(\mathscr{C}) \subseteq \sigma(\mathscr{C}^*)$;

(ii) if $\mathscr{C} \subseteq \mathscr{C}^* \subseteq \sigma(\mathscr{C})$, then

$$\sigma(\mathscr{C}^*) = \sigma(\mathscr{C});$$

(iii) $\quad \sigma\{\sigma(\mathscr{C})\} = \sigma(\mathscr{C})$;

(iv) $\quad \sigma\left(\bigcup_{t \in T} \mathscr{C}_t\right) = \sigma\left\{\bigcup_{t \in T} \sigma(\mathscr{C}_t)\right\}$.

Deduce from (ii) that $\sigma(\mathscr{I}) = \mathscr{B}$ and, more generally, that $\sigma(\mathscr{I}^n) = \mathscr{B}^n$.

Hint. For the last part use Theorem 2 and Exercise 7.

10. (i) Suppose that $A \in \mathscr{B}$. Prove that $A \times R^{n-1}$, i.e.

$$\{(x_1, x_2, \ldots, x_n) : x_1 \in A\},$$

belongs to $\mathscr{B}^n$.

Hint. Let $\mathscr{C}$ be the class of those sets $A \in \mathscr{B}$ for which $A \times R^{n-1} \in \mathscr{B}^n$. Verify that $\mathscr{C}$ is a sigma-algebra and that $\mathscr{C} \supseteq \mathscr{J}$.

Deduce that if $A_1, A_2, \ldots, A_n \in \mathscr{B}$ then $A_1 \times A_2 \times \ldots \times A_n \in \mathscr{B}^n$.

(ii) Suppose that $A \in \mathscr{B}^m$. Prove that $A \times R^n$, i.e.

$$\{(x_1, \ldots, x_m, x_{m+1}, \ldots, x_{m+n}) : (x_1, \ldots, x_m) \in A\},$$

belongs to $\mathscr{B}^{m+n}$.

Deduce that if $A \in \mathscr{B}^m$ and $B \in \mathscr{B}^n$ then $A \times B \in \mathscr{B}^{m+n}$.

(iii) Generalise the result of part (ii) to the Cartesian product of any finite number of Borel sets.

11. Suppose that $f\colon \Omega \to \Omega'$ and $g\colon \Omega' \to \Omega''$, and that

$$h(\omega) = g\{f(\omega)\} \quad (\omega \in \Omega)$$

($h = g \circ f$). Prove that

$$h^{-1}(A'') = f^{-1}\{g^{-1}(A'')\} \quad (A'' \subseteq \Omega'').$$

12. Suppose that $f\colon \Omega \to \Omega'$, and let $\mathscr{C}'$ be a semi-algebra of sets in Ω'. Prove that $f^{-1}(\mathscr{C}')$ is a semi-algebra in Ω.

13. Let A, B be any two sets in Ω. Show that, at each point of Ω,

$$I_{A \cup B} = I_A + I_B - I_{A \cap B} = \max(I_A, I_B)$$

and

$$I_{A \cap B} = I_A \times I_B = \min(I_A, I_B).$$

Show also that $A \supseteq B$ if and only if $I_A \geqslant I_B$ at each point of Ω.

14. Suppose that $f\colon \Omega \to \Omega'$. For any set $A \subseteq \Omega$ let

$$f(A) = \{f(\omega) : \omega \in A\}$$

(and so $f(A)$ is a set in Ω'). Thus f is now being regarded as a function from $\mathscr{P}(\Omega)$ to $\mathscr{P}(\Omega')$.

(a) Prove the following results:

(i) $f(\emptyset) = \emptyset$, but only $f(\Omega) \subseteq \Omega'$ (there may be strict inclusion)

(ii) If $A_t \subseteq \Omega$ for all t of some index set T, then

$$f\left(\bigcup_{t\in T} A_t\right) = \bigcup_{t\in T} f(A_t),$$

but only

$$f\left(\bigcap_{t\in T} A_t\right) \subseteq \bigcap_{t\in T} f(A_t)$$

(once again there may be strict inclusion).

(iii) If $A \subseteq B \subseteq \Omega$, then $f(A) \subseteq f(B)$.

(iv) Both the assertions

$$f(\Omega - A) \subseteq \Omega' - f(A) \quad (A \subseteq \Omega)$$

and

$$\Omega' - f(A) \subseteq f(\Omega - A) \quad (A \subseteq \Omega)$$

are false.

Note. The mapping $f^{-1}: \mathscr{P}(\Omega') \to \mathscr{P}(\Omega)$ and the operations of forming a union, an intersection or a complement are commutative (see Theorem 3 (ii) and (iii)). The apparently more straightforward mapping $f: \mathscr{P}(\Omega) \to \mathscr{P}(\Omega')$ commutes only with the operation of forming a union.

(b) Show that

(i) $A \subseteq f^{-1}\{f(A)\} \quad (A \subseteq \Omega)$;

(ii) $A' \supseteq f\{f^{-1}(A')\} \quad (A' \subseteq \Omega')$.

Construct examples to show that strict inclusion is possible in each case

(c) For any non-empty class $\mathscr{C}$ of sets in Ω let

$$f(\mathscr{C}) = \{f(A): A \in \mathscr{C}\}$$

(and so $f(\mathscr{C})$ is a non-empty class of sets in Ω'). Construct an example to show that if $\mathscr{F}$ is a sigma-algebra in Ω, then $f(\mathscr{F})$ is not necessarily a sigma-algebra in Ω', even if $f(\Omega) = \Omega'$.

Note. Compare Theorems 3(vii) and 3(viii).

2
Probability

In this chapter we shall first discuss the nature of the experimental evidence which provides the motive for developing probability theory. This evidence will suggest certain axioms (known as Kolmogorov's axioms) which provide a suitable starting point for that theory. After these axioms have been stated, the remainder of the chapter will be devoted to obtaining certain results relating to the probabilities of events which can be deduced from the axioms in a fairly straightforward manner. In doing this, the concepts of conditional probability and independence (of events) will be introduced.

The results obtained in this chapter will, with one exception, be limited to those we shall need later. Consequently many interesting and important theorems (e.g. all theorems relating to what may conveniently be termed "occupancy" problems) receive no mention. This is another reason for carrying out some such preparatory reading as that mentioned in the preface.

The experimental evidence

If an experiment E is performed N times "under identical conditions", and in $N(A)$ of those times a particular event A is observed to occur, then experience suggests that $N(A)/N$ is approximately constant for large N, or that $N(A)/N$ seems to tend to a limit as N tends to infinity. For example, if E consists of tossing a penny on to a hard level surface, and A is the event that the coin falls head uppermost, then $N(A)/N$ is approximately $\frac{1}{2}$ for large N.

The reader will notice that the phrase "under identical conditions" is placed in inverted commas. This is to draw attention to the fact that it is not to be taken literally (if for no other reason than that there must be a change of time between successive performances of the same experiment) but rather in the sense that any conditions which we believe could affect the outcome of the experiment remain unchanged. If, for instance, the hard level surface of the penny-tossing experiment were to be replaced by soft sand, then the possibility that the coin would remain standing on its edge after landing would become appreciable, with a consequent reduction in the approximate value of $N(A)/N$ for large N.

As a further example, let E be the experiment of determining the sex of a live-born child in England and Wales, and let A be the event that the child is a boy. The following table gives the value of $N(A)/N$ for some recent years (N is the total number of live births in the year in question).

Year	1965	1966	1967	1968	1969	1970
$N(A)/N$	0·514	0·515	0·514	0·514	0·514	0·514

The reader is invited to pause here and consider what his reaction would be if he were told that in 1971 the value of $N(A)/N$ had been 0·532. Knowing that there were about 800,000 live births each year in England and Wales, he would first of all expect some error in the registration of the births or in the calculation based on the resulting figures. If this is ruled out, he would seek some cause for so marked a change (e.g. a development in medical science, or massive immigration of some race producing a higher proportion of boys). He would, in effect, no longer believe that the experiment was being performed "under identical conditions".

Thus given the experiment E (and the given conditions), then with the event A is associated a number $P(A)$ (the *probability* of A) with the property that $N(A)/N$ is approximately $P(A)$ for large N. The ratio $N(A)/N$ is called the *relative frequency* of the occurrence of A in the N experiments, and probability theory has its genesis in the observed long-term behaviour of relative frequencies, namely that they are approximately constant for large N, or appear to tend to a limit as N tends to infinity (though the truth of this statement cannot be determined by any finite sequence of experiments). This approach to probability theory is precisely analogous to that to classical dynamics. The latter has its genesis in observations of the behaviour of moving bodies. Just as a lump of matter has associated with it a mass, which may be estimated by weighing the lump in a balance, so an event A has associated with it a probability $P(A)$, which can be estimated by determining $N(A)/N$ for large N. Just as classical dynamics is based on certain axioms (Newton's laws), so probability theory will be based on certain axioms, whose derivation will be our first task.

Note. It should be stressed that $P(A)$ depends not only on the event A but also on the experiment E (and the conditions under which it is performed). For this reason some such notation as $P(A; E)$ might be thought preferable.

However, it is not customary, as the experiment E is usually clear from the context (though a very similar notation is used for the closely related idea of conditional probability; see page 23).

Sample space

Just as the results of elementary classical dynamics can be regarded as applying to idealised point particles (be they of chalk or cheese), so it is convenient to regard the axioms and laws of probability theory as applying to idealised models of experiments.

To construct such a model we introduce a *sample space* appropriate to the particular experiment in question. This is defined to be a non-empty set Ω such that to each conceivable outcome of the experiment corresponds precisely one element ω of Ω. For example, in the penny-tossing experiment, the sample space could be taken to consist of just two points, H(ead) and T(ail); if standing on edge is considered possible then Ω could be taken as $\{H, T, \text{edge}\}$. In the experiment of throwing a die on to a table, Ω could be taken as the set of all points of the form (n, x, y, θ), where n is the number of spots showing on the topmost face of the die (and so n is one of the numbers $1, 2, \ldots, 6$), (x, y) are co-ordinates determining the position of the centroid of the face in contact with the table, and θ is a co-ordinate determining the orientation of the die; if, as is usually the case, we are interested only in the number of spots showing on the topmost face of the die, then the much simpler sample space $\{1, 2, 3, 4, 5, 6\}$ is more appropriate. In the experiment of measuring the height of a man in inches, Ω could be taken as any one of the following:

(i) $\{36, 37, \ldots, 108\}$, if the measurements are made to the nearest inch, and heights less than three feet or greater than nine feet are regarded as impossible;

(ii) $[36, 108]$, if the measurements can be made to any degree of accuracy, but the heights are restricted as in (i);

(iii) $(0, \infty)$ if the measurements can be made to any degree of accuracy, and any positive height is considered possible.

It will be clear that to the same experiment correspond many sample spaces, and the choice of a suitable one may be a matter for thought. In many cases, though, one sample space stands out as being the most appropriate one for the experiment under consideration.

To each "event" (a term still to be defined) A corresponds a set in Ω, which will also be denoted by A, and the event A will be said to *occur* if and only if the outcome $\omega \in A$. For example, in the die-throwing experiment, with sample space $\{1, 2, 3, 4, 5, 6\}$, the event of getting an even number of spots on the topmost face corresponds to the set $\{2, 4, 6\}$. The set corresponding to the event may consist of a single point; for example, in the penny-tossing experiment described above ($\Omega = \{H, T\}$), the event of getting a head corresponds to the set $\{H\}$.

To ways of combining events and making statements about them correspond ways of combining sets in Ω and making statements about them. The following table lists the terminologies in corresponding cases.

Events	*Sets in* Ω
A or B (the event which occurs if and only if at least one of A, B occurs)	$A \cup B$
A and B (the event which occurs if and only if both A and B occur)	$A \cap B$
Not A (the event which occurs if and only if A does not occur)	$A^c = \Omega - A$
The certain event	Ω
The impossible event	$\emptyset$
$A_1, A_2, \ldots, A_n$ are mutually exclusive (i.e. no two of the events can occur simultaneously)	$A_1, A_2, \ldots, A_n$ are disjoint (i.e. $A_i \cap A_j = \emptyset$ for $1 \leqslant i < j \leqslant n$)
$A_1, A_2, \ldots, A_n$ are exhaustive (i.e. at least one of $A_1, A_2, \ldots, A_n$ must occur)	$\bigcup_1^n A_i = \Omega$
B occurs whenever A does	$A \subseteq B$

Sometimes the language of sets will be used and sometimes that of events, whichever is the more convenient. Furthermore, we shall drop the distinction between an outcome and its corresponding sample point, or between an event and its corresponding set. There is no real risk of confusion in doing so.

Events are those sets of conceivable outcomes to which probabilities will be assigned in the model. The class of events must be rich enough to fulfil any demands we are likely to make on it. Thus it must include every set of conceivable outcomes in which we are likely to be interested (the "interesting" sets). Moreover, any "reasonable" combination of events must itself be an event. These rather vague statements will be made precise in the Axioms I below.

However, not every set of points in Ω necessarily corresponds to an event. For example, we are not interested in whether the height of a man in inches falls in a specified non-measurable subset of (50, 60). (The reader unfamiliar with Lebesgue theory needs to know only that a non-measurable set is one which never occurs naturally in the normal processes of analysis.) See also note 2 on page 20.

It will be noticed that the class of events is not determined uniquely by the set Ω of all conceivable outcomes, but depends also on what we consider to be the "interesting" sets. Thus let us consider again the die-throwing experiment with $\Omega = \{1, 2, 3, 4, 5, 6\}$. If we are interested in the actual number of spots showing, then the appropriate class of events is $\mathscr{P}(\Omega)$. If, on the contrary, we are interested only in whether the number of spots showing is odd or even, it might be simpler and more appropriate to take as our class of events $\{\emptyset, \{1, 3, 5\}, \{2, 4, 6\}, \Omega\}$. However, in many cases one particular class of events is clearly the most suitable for the model we are trying to construct.

Kolmogorov's axioms

If "interesting" sets of conceivable outcomes are to be events, and "reasonable" combinations of events are themselves to be events, then the axioms for events must surely imply the following statements:

(1) $\Omega, \emptyset$ are events.
(2) If A is an event, so is A^c.
(3) If A and B are events, so are $A \cup B$ and $A \cap B$.
(4) More generally, if $A_1, A_2, \ldots, A_n$ are events, so are $A_1 \cup A_2 \cup \ldots \ldots \cup A_n$ and $A_1 \cap A_2 \cap \ldots \cap A_n$.

Although (4) suffices for many elementary applications of probability theory, in particular to simple problems of a combinatorial nature, yet to develop the theory it is convenient to extend (4) to infinite sequences of events, and to require also that the following holds:

(5) If $A_1, A_2, \ldots$ are events, so are $\bigcup_1^\infty A_i$ and $\bigcap_1^\infty A_i$.

We select certain of the above statements as the axioms for events, and then show that the other statements follow from these axioms (see Lemma 1).

Axioms I (*for events*) If $\mathscr{F}$ is the class of all events (i.e. a class of sets in Ω), then

(a) $\Omega \in \mathscr{F}$;

(b) if $A \in \mathscr{F}$, then $A^c \in \mathscr{F}$;

(c) if $A_1, A_2, \ldots \in \mathscr{F}$, then $\bigcup_1^\infty A_i \in \mathscr{F}$.

Thus the Axioms I say that $\mathscr{F}$ is a sigma-algebra in Ω. The sests of $\mathscr{F}$ are called *events.*

Lemma 1. *Let $\mathscr{F}$ be the class of all events in Ω. Then*

(i) $\emptyset \in \mathscr{F}$;

(ii) *if* $A_1, A_2, \ldots, A_n \in \mathscr{F}$, *then* $\bigcup_1^n A_i$ *and* $\bigcap_1^n A_i \in \mathscr{F}$;

(iii) *if* $A_1, A_2, \ldots \in \mathscr{F}$, *then* $\bigcap_1^\infty A_i \in \mathscr{F}$.

Proof. (i) $\emptyset = \Omega^c \in \mathscr{F}$ (by Axioms I (a) and (b)).

(ii) $\bigcup_1^n A_i = A_1 \cup A_2 \cup \ldots \cup A_n \cup \emptyset \cup \emptyset \cup \ldots \in \mathscr{F}$

(by (i) and Axiom I (c)).

$$\bigcap_1^n A_i = \left(\bigcup_1^n A_i^c\right)^c \in \mathscr{F}$$

(by Axiom I (b) and what has just been established).

(iii) $\bigcap_1^\infty A_i = \left(\bigcup_1^\infty A_i^c\right)^c \in \mathscr{F}$

(by Axioms I (b) and (c)).

Consequently, those of the statement (1)–(5) which were not enshrined in the axioms do, in fact, follow from the axioms. In particular, the omission of (4) from the axioms has not led to the sacrifice of our finite birthright ((4)) for a mess of infinite pottage ((5)).

The second group of axioms deals with the probabilities themselves. These axioms are suggested by properties of relative frequencies; for example:

(1) For any event A the inequalities

$$0 \leqslant N(A) \leqslant N, \quad \text{i.e.} \quad 0 \leqslant N(A)/N \leqslant 1,$$

always hold.

(2) If A is the certain event ($= \Omega$) then $N(A) = N$, i.e. $N(A)/N = 1$, and if A is the impossible event ($= \emptyset$) then $N(A) = 0$, i.e. $N(A)/N = 0$. (Note that the converses of the two statements made here are false.)

(3) If $A_1, A_2, \ldots, A_n$ are mutually exclusive events and $B = \bigcap_1^n A_i$ (and so B is the event which occurs if and only if at least one, and therefore, since $A_1, A_2, \ldots, A_n$ are assumed to be mutually exclusive, precisely one, of the events $A_1, A_2, \ldots, A_n$ occurs) then

$$N(B) = N(A_1)+N(A_2)+ \ldots +N(A_n),$$

and so

$$\frac{N(B)}{N} = \frac{N(A_1)}{N}+\frac{N(A_2)}{N}+ \ldots +\frac{N(A_n)}{N}.$$

The corresponding statements for probabilities are as follows:

(1) For any event A, $0 \leqslant P(A) \leqslant 1$.

(2) $P(\Omega) = 1, \quad P(\emptyset) = 0$.

(3) If $A_1, A_2, \ldots, A_n$ are mutually exclusive events, then

$$P\left(\bigcup_1^n A_i\right) = \sum_1^n P(A_i).$$

Once again, to develop the theory it is convenient to extend (3) to infinite sequences of events, and to require also that the following holds:

(4) If $A_1, A_2, \ldots$ are mutually exclusive events, then

$$P\left(\bigcup_1^\infty A_i\right) = \sum_1^\infty P(A_i).$$

We select certain of the above statements as axioms for probabilities, and then show that the other statements follow from these axioms (see Lemma 2).

Axioms II (*for probabilities*) There exists a function P: $\mathscr{F} \to R$ such that

(a) $P(A) \geqslant 0$ for every $A \in \mathscr{F}$;

(b) $P(\Omega) = 1$;

(c) if $A_1, A_2, \ldots \in \mathscr{F}$ and are mutually exclusive (i.e. $A_i \cap A_j = \emptyset$ whenever $i \neq j$), then

$$P\left(\bigcup_1^\infty A_i\right) = \sum_1^\infty P(A_i). \tag{1}$$

In (c), $\bigcup_1^\infty A_i$ is the event which occurs if and only if at least one, and therefore (since $A_1, A_2, \ldots$ are assumed to be mutually exclusive) precisely one, of the events $A_1, A_2, \ldots$ occurs. Furthermore, the equation (1) implies that $\sum_1^\infty P(A_i)$ is convergent whenever $A_1, A_2, \ldots$ are disjoint sets of $\mathscr{F}$.

The Axioms II assert that P is what is called a *probability measure* on $\mathscr{F}$.

The number $P(A)$ is called the *probability* of the event A, and the ordered triple $(\Omega, \mathscr{F}, P)$ is called a *probability space*.

Notes. (1) Probability spaces *do* exist. For example, let Ω be any non-empty set containing only a finite number, say n, of points, let $\mathscr{F}$ consist of all the (2^n) sets in Ω, i.e. let $\mathscr{F} = \mathscr{P}(\Omega)$, and for any A let

$$P(A) = (\text{number of points in } A)/n.$$

It is easy to verify that $(\Omega, \mathscr{F}, P)$ is a probability space, i.e. that Axioms I and II are satisfied.

Such a probability space provides a suitable model for an experiment which must result in one of a finite number of mutually exclusive and *equally likely* outcomes (corresponding to the points of Ω). This is the case with many elementary games of chance.

(2) It may not be possible for $\mathscr{F}$ to consist of all the sets in Ω. For example, let the experiment in question be the choosing of a number "at random" in the interval (0, 1), and so we take $\Omega = (0, 1)$. To any sub-interval (a, b) of (0, 1) should be assigned the probability $b-a$ (this assignment makes precise the meaning of the phrase "at random"). It can be shown that it is *not* possible to assign a probability to every subset of (0, 1) in such a way that Axioms II hold and $P\{(a, b)\} = b-a \, (0 \leqslant a < b \leqslant 1)$. The method of proof is, in effect, that used to show the existence of non-measurable sets in Lebesgue theory.

For the remainder of the book we shall assume that, unless otherwise stated or implied, we are given a probability space $(\Omega, \mathscr{F}, P)$.

Lemma 2. (i) $P(\emptyset) = 0$.

(ii) *If* $A_1, A_2, \ldots, A_n \in \mathscr{F}$ *and are mutually exclusive, then*

$$P\left(\bigcup_1^n A_i\right) = \sum_1^n P(A_i).$$

(iii) *For any* $A \in \mathscr{F}$, $P(A^c) = 1 - P(A)$.

(iv) *For any* $A \in \mathscr{F}$, $P(A) \leqslant 1$.

(v) *For* $A \in \mathscr{F}$, $B \in \mathscr{F}$ *and* $A \subseteq B$, $P(A) \leqslant P(B)$.

Proof. (i) $P(\emptyset) = P(\emptyset \cup \emptyset \cup \emptyset \cup \ldots)$

$$= P(\emptyset) + P(\emptyset) + P(\emptyset) + \ldots \text{ (by Axiom II(c))}.$$

Therefore, since $P(\emptyset)$ is a real number, and therefore finite, it must be zero.

(ii) $P\left(\bigcup_{1}^{n} A_i\right) = P(A_1 \cup A_2 \cup \ldots \cup A_n \cup \emptyset \cup \emptyset \cup \ldots)$

$= P(A_1)+P(A_2)+ \ldots +P(A_n)+P(\emptyset)+P(\emptyset)+ \ldots$

(by Axiom II (c))

$= \sum_{1}^{n} P(A_i)$ (by (i)).

(iii) $1 = P(\Omega)$ (by Axiom II (b))

$= P(A \cup A^c)$

$= P(A)+P(A^c)$ (by (ii)),

and so $P(A^c) = 1-P(A)$.

(iv) $P(A) = 1-P(A^c)$ (by (iii))

$\leqslant 1$ (because $P(A^c) \geqslant 0$, by Axiom II (a)).

(v) $P(B) = P\{A \cup (B-A)\}$

$= P(A)+P(B-A)$ (by (ii))

$\geqslant P(A)$ (by Axiom II (a)).

Theorem 1 *Suppose that* $A_1, A_2, \ldots \in \mathscr{F}$. *Then*

$$P\left(\bigcup_{1}^{\infty} A_i\right) \leqslant \sum_{1}^{\infty} P(A_i).$$

Note. The series on the right-hand side is not necessarily convergent.

Proof. The first step is to express $\bigcup_{1}^{\infty} A_i$ as the union of a sequence of disjoint sets of $\mathscr{F}$ by means of the standard identity

$$\bigcup_{1}^{\infty} A_i = A_1 \cup (A_1^c \cap A_2) \cup (A_1^c \cap A_2^c \cap A_3) \cup \ldots$$

(justify it). Then

$$P\left(\bigcup_{1}^{\infty} A_i\right) = P(A_1)+P(A_1^c \cap A_2)+P(A_1^c \cap A_2^c \cap A_3)+ \ldots$$

(by Axiom II (c))

$\leqslant P(A_1)+P(A_2)+P(A_3)+ \ldots$ (by Lemma 2 (v)).

Corollary *Suppose that $A_1, A_2, \ldots, A_n \in \mathscr{F}$. Then*

$$P\left(\bigcup_1^n A_i\right) \leqslant \sum_1^n P(A_i).$$

Proof. Left to the reader.

Theorem 2 *Suppose that $A_1, A_2, \ldots \in \mathscr{F}$, and that*

$$\textit{either} \quad \text{(i) } A_1 \subseteq A_2 \subseteq A_3 \subseteq \ldots \quad \textit{and} \quad A = \bigcap_1^\infty A_i$$

$$\textit{or} \quad \text{(ii) } A_1 \supseteq A_2 \supseteq A_3 \supseteq \ldots \quad \textit{and} \quad A = \bigcap_1^\infty A_i.$$

Then $P(A_n) \to P(A)$ as $n \to \infty$.

Proof (i). The set A can be expressed as the union of a sequence of disjoint sets of $\mathscr{F}$ as follows:

$$A = A_1 \cup (A_2 - A_1) \cup (A_3 - A_2) \cup \ldots$$

(this identity is, in fact, a special case of that used in the proof of Theorem 1). Therefore

$$\begin{aligned} P(A) &= P(A_1) + P(A_2 - A_1) + P(A_3 - A_2) + \ldots \qquad \text{(by Axiom II (c))} \\ &= \lim_{n\to\infty} \{P(A_1) + P(A_2 - A_1) + \ldots + P(A_n - A_{n-1})\} \\ &\qquad\qquad \text{(by the definition of convergence of an infinite series)} \\ &= \lim_{n\to\infty} P\{A_1 \cup (A_2 - A_1) \cup \ldots \cup (A_n - A_{n-1})\} \qquad \text{(by Lemma 2 (ii))} \\ &= \lim_{n\to\infty} P(A_n). \end{aligned}$$

(ii) Since $A_1^c \subseteq A_2^c \subseteq A_3^c \subseteq \ldots \quad$ and $\quad A^c = \bigcup_1^\infty A_i^c$,

it follows from (i) that

$$P(A_n^c) \to P(A^c) \quad \text{as} \quad n \to \infty.$$

Therefore, by Lemma 2 (iii),

$$1 - P(A_n) \to 1 - P(A) \quad \text{as} \quad n \to \infty,$$

and the result follows.

Warning. If A is the certain event (Ω) then $P(A) = 1$, and if A is the impossible event ($\emptyset$) then $P(A) = 0$. It must, however, be stressed that the converses of these statements need not be true (though they do hold for those probability spaces defined in note 1 on page 20 which provide a suitable model for an experiment which must result in one of a finite number of mutually exclusive and equally likely outcomes). For example, consider the probability space $(\Omega, \mathscr{F}, P)$ for which Ω consists of precisely two points a and b, $\mathscr{F} = \mathscr{P}(\Omega)$, and

$$P(\emptyset) = P(\{a\}) = 0, \quad P(\Omega) = P(\{b\}) = 1.$$

Then $P(\{a\}) = 0$, but the event $\{a\}$ is not impossible (i.e. $\{a\} \neq \emptyset$). Less formally, we can consider the experiment of choosing a number at random from $(0, 1)$. If, say, $A = \{\frac{1}{2}\}$ (i.e. A is the event that the randomly chosen number is $\frac{1}{2}$), then it is intuitively clear that $P(A) = 0$. However, A is not impossible.

The assertion $P(A) = 0$ can also be discussed in terms of relative frequencies, when it is equivalent to the statement that $N(A)/N$ is nearly 0 for a large number N of performances of the experiment. Thus in any single performance of the experiment it is virtually certain that A will not occur.

Conditional probability

Suppose that in a population of N people, $N(A)$ have mathematical ability and $N(B)$ have musical ability. To answer the question "Are people musical?", one computes $N(B)/N$, which, for large N, is approximately $P(B)$ (the probability that a randomly chosen person has musical ability). To answer, the question "Are mathematicians musical?", one computes $N(A \cap B)/N(A)$, where $N(A \cap B)$ is the number of those who have both abilities. Note that

$$\frac{N(A \cap B)}{N(A)}$$

is, for large N, approximately equal to the probability that a randomly chosen mathematician has musical ability, and that it is also equal to

$$\frac{N(A \cap B)}{N} \bigg/ \frac{N(A)}{N},$$

which, for large N, is approximately $P(A \cap B)/P(A)$.

Thus $P(A \cap B)/P(A)$ is our reassessment of the probability that a randomly chosen person has musical ability in the light of the extra information that he has mathematical ability, and we are led to the following definition.

Definition. Suppose that A is an event with $P(A) > 0$. Then *the conditional probability of an event B given that A has occurred* (or, more shortly, *the* *probability of B given A*) is defined to be

$$P(A \cap B)/P(A), \tag{2}$$

and written $P(B \mid A)$.

The longer form "... given that A has occurred" would seem to imply that the conditional probability $P(B \mid A)$ can be defined only when B follows A in time. Although this is often the case in practice, it is not necessarily so. The shorter form, as being less misleading, is therefore doubly to be preferred.

Independence

The conditional probability $P(B \mid A)$ is the probability of B reassessed in the light of the additional information that A is known to have occurred. It can happen that this additional information may leave our estimate of the probability unaltered, i.e. that $P(B \mid A) = P(B)$.

Consider, in particular, two events A and B which are physically independent. For example, let an experiment consist of the simultaneous tossing of a penny and the throwing of a die, and let A be the event that the penny shows a head, and B the event that the die shows a six. Then the occurrence of A has no effect on that of B, and the proportion of times that the event B occurs in N performances of the experiment should be approximately equal to the proportion of times B occurs in those $N(A)$ experiments in which A occurs, i.e.

$$\frac{N(A \cap B)}{N(A)} \quad \text{and} \quad \frac{N(B)}{N}$$

should be approximately equal. In such a case we should have

$$P(B \mid A) = P(B) \quad \text{or} \quad P(A \cap B) = P(A)\,P(B).$$

The second of these relations suggests the following definition.

Definition. The events A and B are *stochastically independent* if

$$P(A \cap B) = P(A)\,P(B).$$

If, as is usually the case, there can be no confusion with other meanings of the word, one refers simply to *independent* events.

What are, in effect, alternative forms of the above definition are given in Exercise 7.

Extension to more than two events

The idea that the definition of independence formalises is that information about the occurrence of A does not affect the probability of occurrence of B (and, by the symmetry of the definition, vice-versa). In extending the definition of independence to more than two events, the corresponding idea is that information about the occurrence of some of the events does not affect the probability of occurrence of any of the remaining events. For example, with three events A, B and C, not only are

$$\frac{N(A \cap B)}{N(A)} \quad \text{and} \quad \frac{N(B)}{N}$$

to be approximately equal, suggesting (as before) that

$$P(A \cap B) = P(A)P(B) \tag{3}$$

(and two similar equations, involving A and C, B and C), but also

$$\frac{N(A \cap B \cap C)}{N(A \cap B)} \quad \text{and} \quad \frac{N(C)}{N}$$

are to be approximately equal, and so we should expect

$$\begin{aligned} P(A \cap B \cap C) &= P(A \cap B)P(C) \\ &= P(A)P(B)P(C) \quad \text{(in virtue of (3)).} \end{aligned}$$

These considerations generalise to any finite number n of events, and suggest the following definition.

Definition. The n events $A_1, A_2, \ldots, A_n$ are (*stochastically*) *independent* if

$$P(A_{i_1} \cap A_{i_2} \cap \ldots \cap A_{i_r}) = P(A_{i_1})P(A_{i_2}) \ldots P(A_{i_r}) \tag{4}$$

for every $r = 2, 3, \ldots, n$ and, corresponding to each such r, for every choice of r distinct suffixes $i_1, i_2, \ldots, i_r$ from $1, 2, \ldots, n$.

It will be seen that altogether there are

$$\binom{n}{2} + \binom{n}{3} + \ldots + \binom{n}{n} = 2^n - n - 1$$

equations (4).

If only

$$P(A_{i_1} \cap A_{i_2}) = P(A_{i_1})P(A_{i_2}) \tag{5}$$

for every choice of two distinct suffixes i_1, i_2 from 1, 2, ..., n ($\binom{n}{2}$ equations), then $A_1, A_2, \ldots, A_n$ are said to be *pairwise independent*.

If $A_1, A_2, \ldots, A_n$ are independent they are clearly pairwise independent. To show that the converse statement is false, consider the probability space $(\Omega, \mathscr{F}, P)$ for which Ω consists of precisely four points a, b, c and d, $\mathscr{F} = \mathscr{P}(\Omega)$, and for any $A \subseteq \Omega$

$$P(A) = (\text{the number of points in } A)/4.$$

If $A = \{a, d\}$, $B = \{b, d\}$ and $C = \{c, d\}$, then A, B and C are pairwise independent but not independent. For

$$P(A \cap B) = \tfrac{1}{4} = P(A)\,P(B),$$

and two similar equations, but

$$P(A \cap B \cap C) = \tfrac{1}{4} \neq P(A)\,P(B)\,P(C) = \tfrac{1}{8}.$$

It is clear that our intuitive ideas require that if A and B are independent, then so are A^c and B. For if information about the occurrence of A does not affect the probability of occurrence of B, then nor does the precisely equivalent information about the occurrence of A^c. It is easy to give a formal proof.

Let A and B be independent events. Then

$$\begin{aligned} P(A^c \cap B) &= P(B) - P(A \cap B) \quad &&\text{(by Lemma 2 (ii))} \\ &= P(B) - P(A)\,P(B) \quad &&\text{(because } A \text{ and } B \text{ are independent)} \\ &= \{1 - P(A)\}\,P(B) = P(A^c)\,P(B), \end{aligned}$$

and so A^c and B are independent. From this result it then follows that A^c and B^c are independent.

Similarly, if A, B and C are independent events, it can be shown that A^c, B and C are independent, and so on.

Theorem 3 *Let $A_1, A_2, \ldots, A_n$ be independent events, and for each $i = 1, 2, \ldots, n$ let B_i be one of A_i, A_i^c. Then $B_1, B_2, \ldots, B_n$ are independent.*

Proof. It suffices to prove that, say, $A_1^c, A_2, \ldots, A_n$ are independent, and this is left to the reader.

The definition of independence may be extended to the events of any infinite class of events $\{A_t : t \in T\}$, where T is an infinite index set.

Definition. The events of the class $\{A_t : t \in T\}$ are *independent* if the events of every finite sub-class are independent, or equivalently, if

$$P\left(\bigcap_{s=1}^{r} A_{t_s}\right) = \prod_{s=1}^{r} P(A_{t_s})$$

for every $r \geqslant 2$ and, corresponding to each such r, for every choice of r distinct suffixes $t_1, t_2, \ldots, t_r$ from T.

As an introduction to our next theorem, let us consider an idealised experiment, each performance of which consists of an infinite sequence of tosses of an unbiased penny. Then to each outcome of the experiment corresponds an infinite sequence $\omega = (\omega_1, \omega_2, \ldots)$, where $\omega_n = H$ or $= T$ according as the nth toss resulted in a head or a tail. The aggregate of all such sequences is a suitable sample space Ω for this experiment.

An appropriate probability space $(\Omega, \mathscr{F}, P)$ would have the property that any set typified by

$$\{(\omega_1, \omega_2, \ldots) : \omega_1 = H, \quad \omega_2 = T, \quad \omega_3 = T, \ldots, \omega_{n-1} = H, \omega_n = T\}$$

would belong to $\mathscr{F}$, and that the probability measure of any such set would be 2^{-n}. Unfortunately, the proof of the existence of such a probability space is more difficult than the uninitiated might expect, and it will not be given in this book. However, let us assume that this probability space does exist (a result which is intuitively obvious) and continue with the discussion of the experiment. Suppose we ask for the probability that the infinite sequence of tosses contains an infinity of runs of precisely three heads. We can rephrase this question as follows: If A_n is the event that a sequence of precisely three heads commences with the nth toss ($n = 1, 2, \ldots$), what is the probability that an infinity of the events $A_1, A_2, \ldots$ occur?

Let us restate this question in a more general, and more useful, form. Suppose we are given a (general) probability space $(\Omega, \mathscr{F}, P)$ and an infinite sequence of events, i.e. sets of $\mathscr{F}$, and are asked to determine the probability that an infinity of the A_n's occur. A partial answer is given in Theorem 4 below, but the reader should note that a necessary preliminary is to verify that

$$B = \{\omega : \omega \in A_n \text{ for an infinity of values of } n\}$$

is an event, i.e. that $B \in \mathscr{F}$. This was done in Exercise 1.2.

Theorem 4 (*the Borel–Cantelli lemmas*) *Suppose that $A_1, A_2, \ldots$ are events, and let*

$$p_n = P(A_n) \qquad (n = 1, 2, \ldots).$$

Let B be the event which occurs if and only if an infinity of $A_1, A_2, \ldots$ occur. Then

(i) *if $\sum_1^\infty p_n$ is convergent, $P(B) = 0$;*

(ii) *if $\sum_1^\infty p_n$ is divergent and if also $A_1, A_2, \ldots$ are independent, $P(B) = 1$.*

Proof. Let $B_k = \bigcup_{n=k}^{\infty} A_n$ $(k = 1, 2, \ldots)$, and so $B = \bigcap_{k=1}^{\infty} B_k$.

(i) For all k

$$\begin{aligned} 0 \leqslant P(B) &\leqslant P(B_k) \quad \text{(by Lemma 2 (v), because } B \subseteq B_k) \\ &= P\left(\bigcup_{n=k}^{\infty} A_n\right) \\ &\leqslant \sum_{n=k}^{\infty} p_n \quad \text{(by Theorem 1)} \\ &\to 0 \quad \text{as} \quad k \to \infty \quad \left(\text{because } \sum_1^\infty p_n \text{ is convergent}\right). \end{aligned}$$

Therefore $P(B) = 0$.

(ii) It is now more convenient to consider the complementary event

$$B^c = \bigcup_{k=1}^{\infty} B_k^c,$$

where $B_k^c = \bigcap_{n=k}^{\infty} A_n^c$ $(k = 1, 2, \ldots)$.

Let k be a positive integer, and let m be any integer greater than k. Then

$$\begin{aligned} 0 \leqslant P(B_k^c) &= P(A_k^c \cap A_{k+1}^c \cap \ldots) \\ &\leqslant P(A_k^c \cap \ldots \cap A_m^c) \quad \text{(by Lemma 2 (v))} \\ &= P(A_k^c) \ldots P(A_m^c) \quad \text{(by Theorem 3, because } A_k, \ldots, A_m \text{ are independent)} \\ &= \prod_{r=k}^{m} (1-p_r) \\ &\leqslant \exp\left(-\sum_{r=k}^{m} p_r\right), \end{aligned}$$

since $1-x \leqslant e^{-x}$ for all real x. Now, for fixed k,

$$\sum_{r=k}^{m} p_r \to \infty \quad \text{as} \quad m \to \infty$$

(because $\sum_{1}^{\infty} p_n$ is divergent), and so

$$\exp\left(-\sum_{r=k}^{m} p_r\right) \to 0 \quad \text{as} \quad m \to \infty.$$

Therefore $P(B_k^c) = 0$.

This holds for all k, and so, by Theorem 1,

$$P(B^c) = P\left(\bigcup_{k=1}^{\infty} B_k^c\right) \leqslant \sum_{k=1}^{\infty} P(B_k^c) = 0.$$

Therefore $P(B^c) = 0$, and so $P(B) = 1$.

Notes. (1) In (ii) the condition that $A_1, A_2, \ldots$ be independent cannot be omitted. This may be shown by choosing any probability space with an event A for which $0 < P(A) < 1$, and then taking $A_1 = A_2 = \ldots = A$ (and so $B = A$).

(2) If $A_1, A_2, \ldots$ are independent the theorem states that $P(B) = 0$ or 1 according as $\sum_{1}^{\infty} P(A_n)$ is convergent or divergent. This is a particular case of Kolmogorov's zero-one law which, for a sequence of independent events $A_1, A_2, \ldots$, states that any event which is unaffected by the occurrence or otherwise of any finite number of the A's (as is the case with the event B of Theorem 4) must have either probability 0 or probability 1 (see Theorem 6.2).

We can now solve the penny-tossing problem stated above. We note firstly that

$$P(A_n) = 2^{-4} \text{ if } n = 1, \quad \text{and} \quad = 2^{-5} \text{ if } n > 1,$$

and secondly that

$$A_1, A_6, A_{11}, A_{16}, \ldots \tag{6}$$

are independent (note that $A_1, A_2, A_3, \ldots$ are not independent). It follows from Theorem 4(ii) that the probability that an infinity of the events (6) occur is 1, and so the required probability is also 1 (by Lemma 2 (v)).

Completion

This section is devoted to clearing up a theoretical difficulty, but as its results will not be needed later in this book, it may be omitted without impairing the understanding of what follows.

Suppose that $A \in \mathscr{F}$, $P(A) = 0$, and $B \subseteq A$. If $B \in \mathscr{F}$ it follows from Lemma 2(v) that $P(B) = 0$. However, if $B \notin \mathscr{F}$, then $P(B)$ is not defined and no assertion can be made about its value. This is somewhat anomalous, for if A corresponds to some set of outcomes of a hypothetical experiment E, and in N performances of E "under identical conditions" the event A occurs $N(A)$ times, then we expect $N(A)/N$ to be approximately $P(A)$ (i.e. 0) for large N. But $N(B)$ (i.e. the number of times the outcome of E is contained in the set B in the N performances of E) cannot exceed $N(A)$ if $B \subseteq A$, and so $N(B)/N$ is also approximately 0 for large N. Thus B is a set for which $P(B)$ can reasonably be defined (as 0). Consequently, we should like B to belong to $\mathscr{F}$.

Fortunately, the above anomaly is more apparent than real, and can easily be removed. This is done by means of Theorem 5, which requires a preliminary definition.

Definition. The probability space $(\Omega, \mathscr{F}, P)$ is *complete* if whenever $A \in \mathscr{F}$, $P(A) = 0$ and $B \subseteq A$ then $B \in \mathscr{F}$ (and so $P(B) = 0$).

Theorem 5 *Let $(\Omega, \mathscr{F}, P)$ be a probability space. Let $\overline{\mathscr{F}}$ be the class of those sets $A \subseteq \Omega$ to which correspond sets $B, C \in \mathscr{F}$ such that $P(C) = 0$ and*

$$B \subseteq A \subseteq B \cup C,$$

i.e. $\overline{\mathscr{F}}$ is the class of those sets which can be expressed in the form (set of $\mathscr{F}$) $\cup$ (subset of a set of $\mathscr{F}$ having zero probability), and for any such set A let

$$\bar{P}(A) = P(B). \tag{7}$$

Then

(i) *$\overline{\mathscr{F}}$ is a sigma-algebra;*
(ii) *$\overline{\mathscr{F}} \supseteq \mathscr{F}$;*
(iii) *for each $A \in \overline{\mathscr{F}}$, $\bar{P}(A)$ is uniquely defined by* (7) *(i.e. it does not depend on the choice of B);*
(iv) *$\bar{P} = P$ on $\mathscr{F}$;*
(v) *$\bar{P}$ is a probability measure on $\overline{\mathscr{F}}$;*
(vi) *the probability space $(\Omega, \overline{\mathscr{F}}, \bar{P})$ is complete;*
(vii) *if the probability space $(\Omega, \mathscr{F}_1, P_1)$ is complete, $\mathscr{F}_1 \supseteq \mathscr{F}$, and $P_1 = P$ on $\mathscr{F}$, then $\mathscr{F}_1 \supseteq \overline{\mathscr{F}}$ and $P_1 = \bar{P}$ on $\overline{\mathscr{F}}$.*

Note. $(\Omega, \overline{\mathscr{F}}, \bar{P})$ is called the *completion* of $(\Omega, \mathscr{F}, P)$. By (vii), it is the "smallest" complete probability space "containing" $(\Omega, \mathscr{F}, P)$.

Proof. As the proof is straightforward, requiring little more than the definitions of the various terms involved, and as the theorem will not be required later, we content ourselves with proving part (iii), and leaving the remainder of the proof to the reader.

Suppose $A \in \overline{\mathscr{F}}$. Then there exist $B, C \in \mathscr{F}$ such that $P(C) = 0$ and $B \subseteq A \subseteq B \cup C$. Suppose also that there exist $B_1, C_1 \in \mathscr{F}$ such that $P(C_1) = 0$ and $B_1 \subseteq A \subseteq B_1 \cup C_1$. Then

$$B \subseteq B_1 \cup C_1$$

and so

$$\begin{aligned} P(B) &\leqslant P(B_1 \cup C_1) && \text{(by Lemma 2 (v))} \\ &\leqslant P(B_1)+P(C_1) && \text{(by Theorem 1, corollary)} \\ &= P(B_1) && \text{(because } P(C_1) = 0\text{).} \end{aligned}$$

Similarly $P(B_1) \leqslant P(B)$, and so $P(B_1) = P(B)$.

Thus any probability space $(\Omega, \mathscr{F}, P)$ can be extended to a complete probability space $(\Omega, \overline{\mathscr{F}}, \bar{P})$ in such a manner that events in the first space (i.e. sets of $\mathscr{F}$) are events in the second space (i.e. sets of $\overline{\mathscr{F}}$) with the same probability (by Theorem 5 (ii) and (iv)). Since $(\Omega, \overline{\mathscr{F}}, \bar{P})$ is complete, every subset of any set A of $\overline{\mathscr{F}}$ for which $\bar{P}(A) = 0$ is itself a set of $\overline{\mathscr{F}}$, and the anomaly we discussed above has been removed.

Exercises

1. Let A and B be events. Prove that

$$P(A \cup B)+P(A \cap B) = P(A)+P(B).$$

2. Let $A_1, A_2, \ldots$ be events. Prove the following:

(i) If $P(A_1) = P(A_2) = \ldots = 0$

then
$$P\left(\bigcup_{n=1}^{\infty} A_n\right) = 0.$$

Hint. Theorem 1.

(ii) If $P(A_1) = P(A_2) = \ldots = 1$

then
$$P\left(\bigcap_{n=1}^{\infty} A_n\right) = 1.$$

Note. The result of part (i) is a particularly useful one.

3. Let $\mathscr{S} = \{A : A \in \mathscr{F} \text{ and } P(A) = 0 \text{ or } 1\}$. Prove that $\mathscr{S}$ is a sigma-algebra of sets in Ω.

4. Suppose that $P(A_0) > 0$ and, for all $A \in \mathscr{F}$, let $P_0(A)$ be defined to be $P(A \mid A_0)$. Prove that $(\Omega, \mathscr{F}, P_0)$ is a probability space.

5. Let A, B, C be three events, and suppose that $P(A \cap B) > 0$. Prove that

$$P(A \cap B \cap C) = P(A)\, P(B \mid A)\, P(C \mid A \cap B).$$

State and prove the corresponding result for n events.

6. (i) Suppose that the events $A_1, A_2, \ldots, A_n$ are exhaustive and mutually exclusive, and that

$$P(A_i) > 0 \qquad (i = 1, 2, \ldots, n).$$

Prove that, for any event B,

$$P(B) = \sum_{i=1}^{n} P(B \mid A_i)\, P(A_i).$$

(ii) If also $P(B) > 0$, prove that, for $r = 1, 2, \ldots, n$,

$$P(A_r \mid B) = P(B \mid A_r)\, P(A_r) \Big/ \left\{ \sum_{i=1}^{n} P(B \mid A_i)\, P(A_i) \right\}.$$

Note. The result (ii) is known as *Bayes's theorem.*

7. (i) Suppose that $P(A) > 0$. Prove that A and B are independent if and only if $P(B) = P(B \mid A)$.

(ii) Suppose that $P(A) > 0$ and $P(A^c) > 0$. Prove that A and B are independent if and only if $P(B \mid A) = P(B \mid A^c)$.

8. Show that an event A is independent of itself if and only if $P(A)$ is either 0 or 1, and that A is then independent of every event B.

9. Suppose that an event A is independent of each of the three events B, C and $B \cup C$. Prove that A is independent of $B \cap C$.

10. Construct an example of a probability space having three events A, B, C which satisfy

$$P(A \cap B \cap C) = P(A)\, P(B)\, P(C),$$

but which are not independent.

11. Let $\Omega = \{1, 2, \ldots\}$ and $\mathscr{F} = \mathscr{P}(\Omega)$, and let a probability measure P on $\mathscr{F}$ be determined by

$$P(\{n\}) = 1/\{n(n+1)\} \qquad (n = 1, 2, \ldots).$$

Let $A_n = \{n, n+1, \ldots\}$ $(n = 1, 2, \ldots)$, and let B be the event which occurs if and only if A_n occurs for an infinity of values of n. Prove that $\sum_{n=1}^{\infty} P(A_n)$ is divergent, but that $P(B) = 0$.

Why does this not contradict the result of Theorem 4?

12. With the notation of Theorem 4, prove that if there exists a positive number c such that $P(A_n) \geqslant c$ for an infinity of values of n, then $P(B) \geqslant c$.

3 Measurability

A random variable may be defined informally as a real-valued observation, or function of the observations, made in performing an experiment, whose value will (in general) vary from performance to performance of the experiment. For example, the number obtained in throwing a die and the square of that number are both random variables.

Somewhat more formally, a random variable may be defined as any real-valued function X with domain Ω ($X : \Omega \to R$), where $(\Omega, \mathscr{F}, P)$ is a probability space.

However, such a definition would permit any real-valued function defined on Ω, however arbitrary its behaviour, to be regarded as a random variable, whereas, to develop the theory, it is desirable to impose some restriction on the function before it can be so regarded.

The restriction is indicated by considering one of the principal questions asked about a series of experimental observations of the same type, e.g. measurements of the heights of randomly selected British males. This question is, typically, "What proportion of the observations lie below a given value? (above a given value? in a given interval?)" or, equivalently, "What is the probability that an observation lies below a given value? (above a given value? in a given interval?)". The functions considered as random variables must be such that this question can be answered; thus if X is a random variable and I is any interval (finite or infinite) on the real line, we require that the set of points

$$\{\omega : X(\omega) \in I\} = X^{-1}(I) \quad (\subseteq \Omega)$$

must be a set with which a probability can be associated, i.e. it must be a set belonging to the sigma-algebra $\mathscr{F}$ of events. We are thus led to the definition which follows.

Let $\mathscr{F}$ be a sigma-algebra of sets in the space Ω, and let f be a real-valued function defined on Ω, i.e. $f : \Omega \to R$.

Definition. f is $\mathscr{F}$*-measurable* if

$$f^{-1}\{(-\infty, c]\} = \{\omega : f(\omega) \leqslant c\} \in \mathscr{F} \tag{1}$$

for every real number c, or equivalently if

$$f^{-1}(\mathscr{I}) \subseteq \mathscr{F}$$

($\mathscr{I}$ was defined in Theorem 1.2).

Notes. (1) If $(\Omega, \mathscr{F}, P)$ is a probability space, a random variable on $(\Omega, \mathscr{F}, P)$ can now be defined as an $\mathscr{F}$-measurable function. Thus random variables are functions measurable with respect to a sigma-algebra on which a probability measure is assumed to exist.

With this definition, a random variable is defined everywhere on Ω, and its values are real numbers. It is sometimes convenient to extend the definition to include functions which are defined almost surely on Ω (a function is said to be defined almost surely on Ω if it is defined at all points of Ω except possibly on some set A for which $P(A) = 0$), and which are allowed to take the values $+\infty$ and $-\infty$. Such an extension of the definition will not be needed in this book, though we shall consider later random variables whose values are complex numbers.

Random variables will be more fully considered in Chapter 4. In this chapter we shall confine our attention to the properties of measurable functions.

(2) When the sigma-algebra is clear from the context, we shall refer to a "measurable" function instead of to an "$\mathscr{F}$-measurable" one.

Further, when in any given context we state that, say, f and g are measurable, this will mean that they are both measurable with respect to the *same* sigma-algebra (unless clearly stated or implied otherwise).

(3) The abbreviated notation $\{f \leqslant c\}$ will often be used instead of $\{\omega : f(\omega) \leqslant c\}$.

(4) The condition (1) of the definition of measurability can be replaced by

$$f^{-1}\{(-\infty, c)\} \in \mathscr{F}$$

or by

$$f^{-1}\{(c, \infty)\} \in \mathscr{F}$$

or by

$$f^{-1}\{[c, \infty)\} \in \mathscr{F}.$$

The proof of this assertion is left to the reader. In addition to the obvious identities

$$f^{-1}\{(c, \infty)\} = \Omega - f^{-1}\{(-\infty, c]\}$$

and

$$f^{-1}\{[c, \infty)\} = \Omega - f^{-1}\{(-\infty, c)\},$$

he will also need the following two slightly less obvious ones (which he should first establish):

$$\{f < c\} = \bigcup_{n=1}^{\infty} \left\{f \leqslant c - \frac{1}{n}\right\}$$

and

$$\{f \leqslant c\} = \bigcap_{n=1}^{\infty} \left\{f < c + \frac{1}{n}\right\}.$$

If f is measurable, the class of sets whose inverse images under f belong to $\mathscr{F}$ is very much wider than that immediately suggested by the definition.

Theorem 1 *Suppose that $f: \Omega \to R$. Then f is measurable if and only if $f^{-1}(B) \in \mathscr{F}$ for every Borel set B in R, i.e. if and only if*

$$f^{-1}(\mathscr{B}) \subseteq \mathscr{F}. \tag{2}$$

Proof. For every real number c, $(-\infty, c]$ is a closed set, and is therefore in $\mathscr{B}$. Consequently, if (2) holds, $f^{-1}\{(-\infty, c]\} \in \mathscr{F}$ for every real number c, and so f is measurable.

Conversely, suppose that f is measurable. Then $f^{-1}(\mathscr{I}) \subseteq \mathscr{F}$, and

$$\begin{aligned} f^{-1}(\mathscr{B}) &= f^{-1}\{\sigma(\mathscr{I})\} \quad &&\text{(by Theorem 1.2)} \\ &= \sigma\{f^{-1}(\mathscr{I})\} \quad &&\text{(by Theorem 1.3 (ix))} \\ &\subseteq \mathscr{F} \end{aligned}$$

(because $\mathscr{F}$ is a sigma-algebra containing all the sets of $f^{-1}(\mathscr{I})$). Therefore (2) holds.

Theorem 2 (i) *Suppose that the function f: $R^n \to R$ is continuous on R^n. Then f is $\mathscr{B}^n$-measurable.*

(ii) *Suppose that the function f: $R \to R$ is monotonic on R. Then f is $\mathscr{B}$-measurable.*

Note. When $\Omega = R^n$ and $\mathscr{F} = \mathscr{B}^n$, an $\mathscr{F}$-measurable function is referred to as *Borel measurable*. Any function f: $R^n \to R$ of the kind which occurs naturally in analysis is Borel measurable.

Proof. (i) Let c be any real number. Since f is continuous, $\{f \leqslant c\}$ is a closed set, and so is in $\mathscr{B}^n$. Therefore f is measurable.

(ii) We may assume that f is non-decreasing. Then it is easy to show that, for any real number c, $\{f \leqslant c\}$ is $\emptyset$, or R, or a set of the form $(-\infty, a)$, or a set of the form $(-\infty, a]$, and so $\{f \leqslant c\}$ is in $\mathscr{B}$. Therefore f is measurable.

The next two theorems show that algebraic combinations of measurable functions are themselves measurable.

Theorem 3 *Suppose that* $f: \Omega \to R$ *is measurable. Then*

(i) *for any real number* k, $f+k$ *and* kf *are measurable;*

(ii) f^2, $|f|$,

$$f^+ = \tfrac{1}{2}(|f|+f) = \begin{cases} f & \textit{if} \quad f \geqslant 0 \\ 0 & \textit{if} \quad f \leqslant 0 \end{cases}$$

and

$$f^- = \tfrac{1}{2}(|f|-f) = \begin{cases} 0 & \textit{if} \quad f \geqslant 0 \\ -f & \textit{if} \quad f \leqslant 0 \end{cases}$$

are measurable;

(iii) *if* $f \neq 0$ *on* Ω, $1/f$ *is measurable;*

(iv) *if* $f \geqslant 0$ *on* Ω, $\sqrt{f}$ *is measurable.*

Notes. (1) By, say, $f+k$ is meant the function $g: \Omega \to R$ defined by

$$g(\omega) = f(\omega)+k \quad (\omega \in \Omega).$$

(2) The reader who is not familiar with the functions f^+ and f^- should sketch the curves $y = f^+(x)$ and $y = f^-(x)$ for, say, $f(x) = \sin x$.

Proof. The proofs follow almost immediately from the definition of measurability. For example (c being any real number)

$$\{f^2 \leqslant c\} = \begin{cases} \emptyset & \text{if} \quad c < 0 \\ f^{-1}(\{0\}) & \text{if} \quad c = 0 \\ f^{-1}\{[-\sqrt{c}, \sqrt{c}]\} & \text{if} \quad c > 0 \end{cases}$$

and

$$\{1/f \leqslant c\} = \begin{cases} f^{-1}\{[1/c, 0)\} & \text{if} \quad c < 0 \\ f^{-1}\{(-\infty, 0)\} & \text{if} \quad c = 0 \\ f^{-1}\{(-\infty, 0)\} \cup f^{-1}\{[1/c, \infty)\} & \text{if} \quad c > 0, \end{cases}$$

and all the sets on the right belong to $\mathscr{F}$ (because f is measurable).

Theorem 4 *Suppose that* $f: \Omega \to R$ *and* $g: \Omega \to R$ *are measurable. Then*

(i) $f+g$ *is measurable;*

(ii) fg *is measurable;*

(iii) $\max(f, g)$ *and* $\min(f, g)$ *are measurable.*

Note. By, say, $\max(f, g)$ is meant the function $h: \Omega \to R$ defined by

$$h(\omega) = \max\{f(\omega), g(\omega)\} \quad (\omega \in R).$$

Proof. (i) It will be shown that $\{f+g < c\} \in \mathscr{F}$ for every real number c, which suffices to establish the measurability of $f+g$ (see note (4) following the definition of measurability).

If $f(\omega)+g(\omega) < c$ then $f(\omega) < c-g(\omega)$, and so there exists a rational number r such that

$$f(\omega) < r < c-g(\omega).$$

Conversely, if such a number r exists, then $f(\omega)+g(\omega) < c$. Therefore

$$\{f+g < c\} = \bigcup_r \{f < r < c-g\},$$

where the union is taken over all rational numbers r, and so

$$\begin{aligned}\{f+g < c\} &= \bigcup_r [\{f < r\} \cap \{r < c-g\}] \\ &= \bigcup_r [\{f < r\} \cap \{g < c-r\}].\end{aligned}$$

Now, for all rational numbers r, $\{f < r\} \in \mathscr{F}$ and $\{g < c-r\} \in \mathscr{F}$ (because f and g are assumed to be measurable), and so

$$\{f < r\} \cap \{g < c-r\} \in \mathscr{F}.$$

Therefore
$$\bigcup_r [\{f < r\} \cap \{g < c-r\}] \in \mathscr{F}$$

(by the definition of a sigma-algebra, using the fact that the rational numbers are enumerable).

(ii) The proof follows from previous results and the identity

$$fg = \tfrac{1}{4}\{(f+g)^2-(f-g)^2\}.$$

(iii) The proof follows from the identities

$$\{\max(f, g) \leqslant c\} = \{f \leqslant c\} \cap \{g \leqslant c\}$$

and

$$\{\min(f, g) \leqslant c\} = \{f \leqslant c\} \cup \{g \leqslant c\},$$

which hold for any real number c.

Borel measurable functions of measurable functions are themselves measurable. More precisely, we have the following theorem.

Theorem 5 *Suppose that $f: \Omega \to R$ is $\mathscr{F}$-measurable and $g: R \to R$ is $\mathscr{B}$-measurable. Then $g \circ f$ is $\mathscr{F}$-measurable.*

Notes. (1) $g \circ f$ is the function defined on Ω by

$$(g \circ f)(\omega) = g\{f(\omega)\} \quad (\omega \in \Omega).$$

(2) In this theorem, measurability with respect to two sigma-algebras is involved: $\mathscr{F}$ (for functions defined on Ω) and $\mathscr{B}$ (for functions defined on R).

Proof. Let B be any Borel set in R. Then

$$(g \circ f)^{-1}(B) = f^{-1}\{g^{-1}(B)\} \quad \text{(by Exercise 1.11).}$$

Now, by Theorem 1, $g^{-1}(B) \in \mathscr{B}$ (because g is Borel measurable) and so $f^{-1}\{g^{-1}(B)\} \in \mathscr{F}$ (because f is $\mathscr{F}$-measurable). The result now follows.

Theorem 6 *Suppose that*

(a) $f_n: \Omega \to R$ *is measurable* ($n = 1, 2, \ldots$);
(b) *for each $\omega \in \Omega$ the sequence.*

$$f_1(\omega), f_2(\omega), \ldots$$

is bounded. Then

(i) $\sup f_n$ *and* $\inf f_n$ *are measurable;*
(ii) $\lim \sup f_n$ *and* $\lim \inf f_n$ *are measurable;*
(iii) *if also $f_n(\omega)$ tends to a limit $f(\omega)$ as $n \to \infty$ for every $\omega \in \Omega$, f is measurable.*

Notes. (1) Condition (b) asserts that to each $\omega \in \Omega$ correspond real numbers $m(\omega)$ and $M(\omega)$ such that

$$m(\omega) \leqslant f_n(\omega) \leqslant M(\omega) \qquad (n = 1, 2, \ldots).$$

It does not follow that any of the functions f_n is bounded on Ω.

(2) By $\sup f_n$ is meant the function defined on Ω by

$$(\sup f_n)(\omega) = \sup\{f_n(\omega): n = 1, 2, \ldots\} \quad (\omega \in \Omega),$$

i.e. its value at the point ω of Ω is the supremum of the numbers $f_1(\omega)$, $f_2(\omega)$, ... By condition (b), $\sup f_n$ is finite on Ω.

Corresponding remarks apply to the other functions introduced in (i) and (ii).

(3) The conclusion that the limit of a convergent sequence of measurable functions is itself measurable is particularly important. The corresponding

conclusion with "measurable" replaced by "continuous" is false, as may be seen by considering the functions $f_n\colon R \to R$ defined by

$$f_n(x) = (1+nx^2)^{-1} \quad (x \in R).$$

(4) If $f_n(\omega)$ tends to a limit $f(\omega)$ as $n \to \infty$ $(\omega \in \Omega)$, the condition (b) is automatically satisfied, and does not require separate verification. The need for condition (b) arises because we confine our attention to finite-valued functions, and its effect is to ensure that the functions (lim) sup f_n and (lim) inf f_n are finite-valued. If we admit functions which can take infinite values, and make the appropriate change in the definition of measurability, then there is no need for condition (b) in the enunciation of the theorem.

Proof. (i) The proof follows at once from the following identities, which hold for any real number c:

$$\{\sup f_n \leqslant c\} = \bigcap_{n=1}^{\infty} \{f_n \leqslant c\}$$

and

$$\{\inf f_n < c\} = \bigcup_{n=1}^{\infty} \{f_n < c\}.$$

The proof of these two identities is left to the reader, who should note that $<$ cannot be replaced by $\leqslant$ in the second of them (why not?). (This was not so in the proof of Theorem 4 (iii).)

(ii) For $n = 1, 2, \ldots$ let

$$g_n(\omega) = \sup \{f_m(\omega) : m = n, n+1, \ldots\} \quad (\omega \in \Omega).$$

Then, by (i), g_n is measurable $(n = 1, 2, \ldots)$. Again by (i), inf g_n is measurable, for if

$$m(\omega) \leqslant f_n(\omega) \leqslant M(\omega) \qquad (n = 1, 2, \ldots)$$

then

$$m(\omega) \leqslant g_n(\omega) \leqslant M(\omega) \qquad (n = 1, 2, \ldots).$$

Since lim sup f_n = inf g_n, lim sup f_n is measurable.

Similarly lim inf f_n is measurable.

(iii) *First proof.* This consists simply of the observation that if $f_n(\omega) \to f(\omega)$ as $n \to \infty$ $(\omega \in \Omega)$, then f = lim sup f_n is measurable by (ii).

Second proof. In view of the importance of this result we give a proof from first principles, which we interpret here as meaning one which requires no knowledge of the theory of upper and lower limits.

For any real number c, $f(\omega) < c$ if and only if there exists a positive integer k such that $f_n(\omega) < c-(1/k)$ for all sufficiently large n. Therefore

$$\{f < c\} = \bigcup_{k=1}^{\infty}\left\{f_n < c-\frac{1}{k} \text{ for all sufficiently large } n\right\}$$
$$= \bigcup_{k=1}^{\infty}\bigcup_{m=1}^{\infty}\left\{f_n < c-\frac{1}{k} \text{ for all } n \geqslant m\right\}$$
$$= \bigcup_{k=1}^{\infty}\bigcup_{m=1}^{\infty}\bigcap_{n=m}^{\infty}\left\{f_n < c-\frac{1}{k}\right\}.$$

Now all the sets $\{f_n < c-(1/k)\}$ belong to $\mathscr{F}$, and so, successively,

$$A_{mk} = \bigcap_{n=m}^{\infty}\left\{f_n < c-\frac{1}{k}\right\} \in \mathscr{F} \qquad (k, m = 1, 2, \ldots),$$

$$B_k = \bigcup_{m=1}^{\infty} A_{mk} \in \mathscr{F} \qquad (k = 1, 2, \ldots),$$

and

$$\{f < c\} = \bigcup_{k=1}^{\infty} B_k \in \mathscr{F}.$$

Therefore f is measurable.

Our final theorem on measurability will show that a non-negative measurable function is the limit of a sequence of measurable functions of a particularly elementary nature (see also Exercise 8). This theorem will play an important part when we come to define the expected value of a non-negative random variable.

We shall first give the requisite definition, and discuss it.

Definition. f: $\Omega \to R$ is *simple* if it takes only a finite number of distinct values.

Let us call any finite class of disjoint sets whose union is Ω a *partition* of Ω, and let us call it an $\mathscr{F}$*-partition* if all the sets belong to $\mathscr{F}$. Then if f is simple there exists a partition $\{A_1, A_2, \ldots, A_m\}$, say, of Ω such that

$$f = c_1 I_{A_1} + c_2 I_{A_2} + \ldots + c_m I_{A_m},$$

where $c_1, c_2, \ldots, c_m \in R$ (I_A is the indicator function of the set A).

If the c's are assumed to be distinct, then the expression of f in the above form is unique, and f is then measurable if and only if all the A's belong to $\mathscr{F}$, i.e. if and only if the partition is an $\mathscr{F}$-partition. However, it is not always convenient to make this assumption, but in any case f is measurable if (though not necessarily only if) all the A's belong to $\mathscr{F}$.

Theorem 7 *Suppose that $f:\Omega \to R$ is measurable and non-negative. Then there exist simple measurable functions $f_1, f_2, \ldots$ such that, for every $\omega \in \Omega$,*

$$0 \leqslant f_1(\omega) \leqslant f_2(\omega) \leqslant \ldots$$

and

$$f_n(\omega) \to f(\omega) \quad \textit{as} \quad n \to \infty.$$

Proof. For $n = 1, 2, \ldots$ let

$$A_{mn} = \{(m-1)2^{-n} \leqslant f < m2^{-n}\} \qquad (m = 1, 2, \ldots, n2^n),$$
$$A_n = \{f \geqslant n\},$$

and

$$f_n = \sum_{m=1}^{n2^n} (m-1)2^{-n} I_{A_{mn}} + nI_{A_n}.$$

(The reader will find the following argument easier to understand if he now sketches the graphs of $f_1(x), f_2(x)$ and $f_3(x)$ for $f(x) = x^2$.)

All the A_{mn}'s and A_n's belong to $\mathscr{F}$, and all the f_n's are simple, measurable and non-negative. Furthermore, for $n = 1, 2, \ldots,$

$$f_n(\omega) \leqslant f_{n+1}(\omega) \quad (\omega \in \Omega).$$

For suppose $\omega \in A_{mn}$; then

$$(m-1)2^{-n} \leqslant f(\omega) < m2^{-n}.$$

Therefore $\quad (2m-2)2^{-n-1} \leqslant f(\omega) < 2m2^{-n-1}.$

Therefore either $\quad f_{n+1}(\omega) = (2m-2)2^{-n-1} = f_n(\omega)$

or $\quad f_{n+1}(\omega) = (2m-1)2^{-n-1} = f_n(\omega) + 2^{-n-1}.$

Similarly, if $\omega \in A_n$ then $f_{n+1}(\omega)$ is one of the numbers $f_n(\omega) + r2^{-n-1}$ $(r = 0, 1, \ldots, 2^{n+1}-1)$ if $n \leqslant f(\omega) < n+1$, or $f_n(\omega)+1$ if $f(\omega) \geqslant n+1$.

Finally, if $n > f(\omega)$, then for some m $(= 1, 2, \ldots, n2^n)$ $\omega \in A_{mn}$ and so

$$f_n(\omega) \leqslant f(\omega) < f_n(\omega) + 2^{-n}.$$

Therefore

$$|f_n(\omega) - f(\omega)| < 2^{-n} \qquad (n > f(\omega)),$$

and so

$$f_n(\omega) \to f(\omega) \quad \text{as} \quad n \to \infty.$$

Exercises

1. Suppose that Ω is a space, and that f is any real-valued function defined on Ω. Prove that

   (i) if $\mathscr{F} = \mathscr{P}(\Omega)$, then f is $\mathscr{F}$-measurable;
   (ii) if $\mathscr{F} = \{\emptyset, \Omega\}$, then f is $\mathscr{F}$-measurable if and only if it is a constant.

2. (i) Suppose that $f: R \to R$ is defined by $f(x) = \sqrt{x}$ if $x \geqslant 0$, and $f(x) = 0$ if $x \leqslant 0$. Prove that f is Borel measurable.

(ii) Suppose that $f: R \to R$ is defined by $f(x) = 1/x$ if $x \neq 0$, and $f(x) = 0$ if $x = 0$. Prove that f is Borel measurable.

3. Give an alternative proof of Theorem 3 using Theorem 5.

4. Suppose that $f: R \to R$ is defined by $f(x) = x$ if $|x| < c$, and $f(x) = 0$ if $|x| \geqslant c$, where $c > 0$. Prove that f is Borel measurable.

5. Suppose that $f: \Omega \to R$ is measurable, and that c is a positive real number. Prove that $|f|^c$ is measurable.

6. Give an alternative proof of Theorem 4(iii) using the results

$$\max(f, g) = \tfrac{1}{2}(f+g+|f-g|)$$

and

$$\min(f, g) = \tfrac{1}{2}(f+g-|f-g|).$$

7. (i) Suppose that f is $\mathscr{F}$-measurable, and that $c \in R$. Prove that $\{f = c\} \in \mathscr{F}$.

(ii) Suppose that f and g are $\mathscr{F}$-measurable. Prove that $\{f = g\}$ and $\{f > g\}$ both belong to $\mathscr{F}$.

8. Suppose that the measurable function f is bounded on Ω. Prove that there exist simple measurable functions $f_1, f_2, \ldots$ such that $f_n \to f$ *uniformly* on Ω as $n \to \infty$.

Hint. There is no loss of generality in assuming that $0 \leqslant f < 1$ on Ω. Define f_n to be $(r-1)/n$ if $(r-1)/n \leqslant f < r/n$ $(r = 1, 2, \ldots, n)$.

9. Let T be an index set, not necessarily countable, and suppose that to each $t \in T$ corresponds a function $f_t: R \to R$ which is continuous on R and satisfies $|f_t| \leqslant 1$ on R. Prove that $\sup f_t$ and $\inf f_t$ are Borel measurable on R. (Note that Theorem 6 is not applicable.)

Hint. For each real number c

$$\{\sup f_t > c\} = \bigcup_{t \in T} \{f_t > c\} \quad \text{is open.}$$

10. The "vector-function"

$$\mathbf{f} = (f_1, f_2, \ldots, f_n): \Omega \to R^n$$

is said to be ($\mathscr{F}$-) measurable if

$$\mathbf{f}^{-1}(\mathscr{I}^n) \subseteq \mathscr{F}.$$

Prove that

(i) $\mathbf{f}$ is measurable if and only if $\mathbf{f}^{-1}(\mathscr{B}^n) \subseteq \mathscr{F}$;
(ii) $\mathbf{f}$ is measurable if and only if $f_1, f_2, \ldots, f_n$ are measurable.

Hints

(i) Proceed as in the proof of Theorem 1, using Exercise 1.7 instead of Theorem 1.2.

(ii) ("if") $\mathbf{f}^{-1}(J_1 \times J_2 \times \ldots \times J_n) = \bigcap_{i=1}^{n} f_i^{-1}(J_i) \quad (J_1, J_2, \ldots, J_n \in \mathscr{I})$.

("only if") $f_1^{-1}(J) = \mathbf{f}^{-1}(J \times R^{n-1}) \quad (J \in \mathscr{I})$.

11. Suppose that $\mathbf{f} : \Omega \to R^n$ is $\mathscr{F}$-measurable and $g : R^n \to R$ is Borel measurable. Prove that $g \circ \mathbf{f} : \Omega \to R$ is $\mathscr{F}$-measurable.

4

Distributions and distribution functions

Suppose that a space Ω and a sigma-algebra $\mathscr{F}$ of sets in Ω are given. Then the $\mathscr{F}$-measurability of a real-valued function defined on Ω depends only on the function and on $\mathscr{F}$, and not on any numerical values which may be assigned to the sets of $\mathscr{F}$. We stated earlier (see page 34) that if a probability measure P is defined on the sets of $\mathscr{F}$, i.e. if $(\Omega, \mathscr{F}, P)$ is a probability space, then a measurable function is called a random variable. For convenience and emphasis, we restate it here in the following formal definition:

Suppose that $(\Omega, \mathscr{F}, P)$ is a probability space, and that $X : \Omega \to R$.

Definition. X is a *random variable* (on the probability space $(\Omega, \mathscr{F}, P)$) if X is $\mathscr{F}$-measurable.

It is customary to use the letters X, Y, Z (with or without affixes) for random variables.

Let X be a given random variable. For any set B of $\mathscr{B}$ let

$$P_X(B) = P\{X^{-1}(B)\}. \tag{1}$$

Since $X^{-1}(B) \in \mathscr{F}$ (by Theorem 3.1), (1) defines a function $P_X : \mathscr{B} \to [0, 1]$. It is left to the reader to show that P_X is a probability measure on $\mathscr{B}$ (see Exercise 1 (i)), and hence that $(R, \mathscr{B}, P_X)$ is a probability space.

The probability measure P_X on $\mathscr{B}$ is called the *distribution* of X. Its significance lies in the fact that any question relating to the random variable X which is of interest in probability theory (for example, that of its expected value – see Chapter 5) can, in principle, be answered when the distribution P_X of X is known. In fact, once the idea of distribution has been extended from a single random variable to a family of random variables (see, for example, Exercise 1(ii)), probability theory can be defined, albeit somewhat uninformatively, as that part of the theory of measurable functions on probability spaces which depends only on the distributions of the functions involved.

A convenient analytical tool for dealing with the distribution of a random variable X is its *distribution function* $F: R \to [0, 1]$ defined for all real x by

$$F(x) = P(X \leqslant x) = P[X^{-1}\{(-\infty, x]\}] = P_X\{(-\infty, x]\}. \qquad (2)$$

($P(X \leqslant x)$ will usually be written in place of what is, strictly speaking, the more correct form $P(\{X \leqslant x\})$; corresponding remarks apply to such expressions as $P(X < x)$ and $P(a < X < b)$.) The distribution function F is, as its name implies, a function; this is also true of the distribution P_X (the terminology is customary, though possibly misleading). The former ($F: R \to [0, 1]$) is a point function, the latter ($P_X : \mathscr{B} \to [0, 1]$) is a set function, i.e. one whose domain is a class of sets. It is clear that F is essentially the restriction of P_X to $\mathscr{I}$, and so the distribution determines the distribution function. It will be shown later that the converse statement is also true, namely that the distribution function determines the distribution. Consequently, any probability theory question relating to a random variable X can, in principle, be answered when its distribution function is known.

There is, of course, no essential difference between set and point functions. The sets in R can be regarded as the points of $\mathscr{P}(R)$; conversely, the points in R can be regarded as the one-point sets in R. However, point functions are usually more familiar and easier to handle, and so the introduction of distribution functions can be expected to be beneficial.

We give two examples to illustrate the above ideas. Firstly, let us consider a binomial $(2, \frac{1}{3})$ random variable; such a random variable has as its possible values 0, 1, 2, with corresponding probabilities $\frac{4}{9}, \frac{4}{9}, \frac{1}{9}$. Then for any Borel set B

$$P_X(B) = \tfrac{4}{9}I_B(0)+\tfrac{4}{9}I_B(1)+\tfrac{1}{9}I_B(2),$$

and for any real number x

$$F(x) = P_X\{(-\infty, x]\} = \begin{cases} 0 & \text{if} \quad x < 0 \\ \frac{4}{9} & \text{if} \quad 0 \leqslant x < 1 \\ \frac{8}{9} & \text{if} \quad 1 \leqslant x < 2 \\ 1 & \text{if} \quad x \geqslant 2. \end{cases}$$

Secondly, let us consider a standardised normal or normal (0, 1) random variable; such a random variable has a probability density function

$$\varphi(x) = (2\pi)^{-\frac{1}{2}} \exp\left(-\tfrac{1}{2}x^2\right) \quad (x \in R),$$

that is to say, a function with the property that

$$P(a < X < b) = \int_a^b \varphi(t)\, dt \quad (-\infty \leqslant a < b \leqslant \infty);$$

if a or b is finite, the corresponding $<$ sign on the left-hand side can be replaced by $\leqslant$. In particular, for every real number x

$$F(x) = P(-\infty < X \leqslant x) = \int_{-\infty}^{x} \varphi(t)\, dt.$$

It can be shown that, for any Borel set B,

$$P_X(B) = \int_B \varphi(t)\, dt,$$

where the integral is to be interpreted as a Lebesgue integral.

Theorem 1 *Suppose that X is a random variable, and that F is the distribution function of X. Then*

(i) *F is non-decreasing on R;*
(ii) *F is everywhere continuous on the right;*
(iii) *$F(x) \to 0$ as $x \to -\infty$ and $F(x) \to 1$ as $x \to \infty$.*

Notes. (1) The distribution function F of a random variable X is sometimes defined by $F(x) = P(X < x)$. If it is so defined, (i) and (iii) still hold, but F is now everywhere continuous on the left.

However, in this book, the distribution function will always be defined by (2).

(2) It will be shown later that any function F having the properties (i), (ii) and (iii) is the distribution function of some random variable (see the corollary to Theorem 3).

Proof. (i) Let a and b be any two real numbers with $a < b$. Then

$$F(b) - F(a) = P(X \leqslant b) - P(X \leqslant a) = P(a < X \leqslant b) \geqslant 0.$$

Therefore F is non-decreasing on R.

(ii) In view of what has just been established, it suffices to prove that

$$F\left(a + \frac{1}{n}\right) \to F(a) \quad \text{as} \quad n \to \infty$$

for every real number a. Now

$$F\left(a + \frac{1}{n}\right) - F(a) = P\left(a < X \leqslant a + \frac{1}{n}\right) = P(A_n),$$

where

$$A_n = \left\{a < X \leqslant a + \frac{1}{n}\right\} \quad (\in \mathscr{F}) \qquad (n = 1, 2, \ldots).$$

Since

$$A_1 \supseteq A_2 \supseteq \dots \quad \text{and} \quad \bigcap_1^\infty A_n = \emptyset,$$

it follows from Theorem 2.2 that

$$P(A_n) \to P(\emptyset) = 0 \quad \text{as} \quad n \to \infty.$$

This completes the proof of (ii).

(iii) It now suffices to prove that $F(-n) \to 0$ and $F(n) \to 1$ as $n \to \infty$. Let

$$B_n = \{X \leqslant n\} \qquad (n = 0, \pm 1, \pm 2, \dots).$$

Then $B_{-1} \supseteq B_{-2} \supseteq \dots$ and $\bigcap_1^\infty B_{-n} = \emptyset$, and so

$$F(-n) = P(B_{-n}) \to P(\emptyset) = 0 \quad \text{as} \quad n \to \infty.$$

Also

$$B_1 \subseteq B_2 \subseteq \dots \quad \text{and} \quad \bigcup_1^\infty B_n = \Omega, \quad \text{and so}$$

$$F(n) = P(B_n) \to P(\Omega) = 1 \quad \text{as} \quad n \to \infty.$$

This completes the proof of (iii).

It will often happen that if $\mathscr{S}$ is a semi-algebra of sets in Ω then a function $\mu : \mathscr{S} \to R$ is, effectively, a probability measure on $\mathscr{S}$. By this is meant that the following statements hold:

(a) $\mu(A) \geqslant 0$ for every $A \in \mathscr{S}$;
(b) $\mu(\Omega) = 1$;
(c) if $A_1, A_2, \dots$ are disjoint and $\in \mathscr{S}$, and also $\bigcup_1^\infty A_n \in \mathscr{S}$, then

$$\mu\left(\bigcup_1^\infty A_n\right) = \sum_1^\infty \mu(A_n)$$

(i.e. μ is "countably additive" on $\mathscr{S}$).

These statements should be compared with Kolmogorov's axioms for probabilities. It should be noted that in (c) it is necessary to require that $\bigcup_1^\infty A_n \in \mathscr{S}$. The corresponding requirement is not necessary in Kolmogorov's axioms, since $\mathscr{F}$ is a sigma-algebra and so automatically $\bigcup_1^\infty A_n \in \mathscr{F}$ if $A_1, A_2, \dots \in \mathscr{F}$. As in the proof of Lemma 2.2, it can be shown that $\mu(\emptyset) = 0$ and also (by a slight modification of the proof there given) that $\mu(A) \leqslant 1$ for every $A \in \mathscr{S}$. (For further properties of μ see Lemma B.1.)

The following theorem asserts that μ can be extended to a probability measure on the sigma-algebra $\sigma(\mathscr{S})$ generated by $\mathscr{S}$.

Theorem 2 (*the extension theorem*) *Let $\mathscr{S}$ be a semi-algebra of sets in Ω, and let $\mu : \mathscr{S} \to R$ satisfy the following conditions:*

(a) $\mu(A) \geqslant 0$ *for every* $A \in \mathscr{S}$;

(b) $\mu(\Omega) = 1$;

(c) *if $A_1, A_2, \ldots$ are disjoint and $\in \mathscr{S}$, and also $\bigcup_1^\infty A_n \in \mathscr{S}$, then*

$$\mu\left(\bigcup_1^\infty A_n\right) = \sum_1^\infty \mu(A_n).$$

Then there exists one and only one probability measure P on $\sigma(\mathscr{S})$ which is equal to μ on $\mathscr{S}$, i.e. for which

$$P(A) = \mu(A) \quad \textit{for every} \quad A \in \mathscr{S}.$$

The extension theorem is a standard result of measure theory. Furthermore, the proof is lengthy, and the details thereof will not be required in what follows. For the sake of completeness, however, we give the proof in Appendix B, to which the interested reader is referred.

Corollary 1 *Let P_1 and P_2 be two probability measures on $\mathscr{B}$ which are equal on $\mathscr{I}$. Then P_1 and P_2 are equal on $\mathscr{B}$.*

Proof. This is an immediate consequence of the uniqueness part of Theorem 2 with $\mathscr{S} = \mathscr{I}$ (and so $\sigma(\mathscr{S}) = \mathscr{B}$, by Exercise 1.9) and μ defined by

$$\mu(A) = P_1(A) = P_2(A) \quad (A \in \mathscr{I}).$$

Corollary 2 *Let P_1 and P_2 be probability measures on $\mathscr{B}$ with corresponding distribution functions F_1 and F_2, i.e.*

$$F_i(x) = P_i\{(-\infty, x]\} \qquad (i = 1, 2;\ x \in R).$$

Then $F_1 = F_2$ on R if and only if $P_1 = P_2$ on $\mathscr{B}$.

Proof. The "if" part of the assertion is immediate. To prove the "only if" part, we note first that the statement

$$F_1(x) = F_2(x) \quad (x \in R)$$

is equivalent to

$$P_1(A) = P_2(A) \quad (A \in \mathscr{I}).$$

Since
$$(a, b] = (-\infty, b] - (-\infty, a]$$
and
$$(a, \infty) = R - (-\infty, a],$$
it follows that
$$P_1(A) = P_2(A) \quad (A \in \mathscr{I}).$$

The conclusion now follows by Corollary 1.

Thus, as stated earlier, the distribution function of a random variable determines its distribution.

If X and Y are any two random variables with corresponding distributions P_X and P_Y, then X and Y are said to be *identically distributed* if $P_X = P_Y$ on $\mathscr{B}$. It follows from Corollary 2 that X and Y are identically distributed if and only if their distribution functions coincide on R.

The next theorem shows that any function F having the properties (i), (ii) and (iii) of Theorem 1 determines a unique probability measure on $\mathscr{B}$ having F as its distribution function.

Theorem 3 *Let $F : R \to [0, 1]$ have the following properties*:

(i) *F is non-decreasing on R*;
(ii) *F is everywhere continuous on the right*;
(iii) *$F(x) \to 0$ as $x \to -\infty$ and $F(x) \to 1$ as $x \to \infty$.*

Then there exists one and only one probability measure P on $\mathscr{B}$ such that
$$P\{(-\infty, x]\} = F(x) \quad (x \in R). \tag{3}$$

Proof. The uniqueness part of the theorem follows immediately from Theorem 2, corollary 2. Thus it remains to prove that there exists at least one probability measure P on $\mathscr{B}$ for which (3) holds.

Let μ be defined on the semi-algebra $\mathscr{I}$ as follows:

$$\mu(\emptyset) = 0,$$
$$\mu\{(a, b]\} = F(b) - F(a) \quad (a, b \in R;\ a < b),$$
$$\mu\{(-\infty, b]\} = F(b) \quad (b \in R),$$
$$\mu\{(a, \infty)\} = 1 - F(a) \quad (a \in R),$$
and
$$\mu(R) = 1.$$

Then $\mu : \mathscr{I} \to R$, and conditions (a) and (b) of Theorem 2 (with Ω, $\mathscr{S}$ replaced by R, $\mathscr{I}$ respectively) are satisfied. Once it has been shown that
$$\mu\left(\bigcup_1^\infty A_n\right) = \sum_1^\infty \mu(A_n)$$

whenever $A_1, A_2, \ldots$ are disjoint and $A_1, A_2, \ldots, \bigcup_1^\infty A_n \in \mathscr{I}$, Theorem 2 can then be applied to establish the existence of a probability measure P on $\mathscr{B}$ which is equal to μ on $\mathscr{I}$, and so, in particular, satisfies (3).

Assume, therefore, that $A_1, A_2, \ldots$ are disjoint sets of $\mathscr{I}$ for which $A = \bigcup_1^\infty A_n \in \mathscr{I}$, and suppose that $A = (a, b]$ (it will be left to the reader to make the necessary modifications to the proof if A is not of this form), and that $A_n = (a_n, b_n]$ $(n = 1, 2, \ldots)$. Then for any positive integer n

$$\mu(A_1)+\mu(A_2)+ \ldots +\mu(A_n) = \sum_{k=1}^{n} \{F(b_k)-F(a_k)\}.$$

Now imagine that the intervals $A_1, A_2, \ldots, A_n$ are relabelled so that in going from left to right along the real line one encounters first A_1, then A_2, and so on, and finally A_n. Then

$$a \leqslant a_1,\ b_1 \leqslant a_2,\ b_2 \leqslant a_3,\ \ldots,\ b_{n-1} \leqslant a_n,\ b_n \leqslant b,$$

and so

$$\left.\begin{aligned} &\sum_{k=1}^{n} \{F(b_k)-F(a_k)\} \\ &\leqslant \sum_{k=1}^{n} \{F(b_k)-F(a_k)\}+\{F(a_1)-F(a)\}+\sum_{k=1}^{n-1} \{F(a_{k+1})-F(b_k)\} \\ &+\{F(b)-F(b_n)\} \quad \text{(because } F \text{ is non-decreasing)} \\ &= F(b)-F(a) \end{aligned}\right\} \tag{4}$$

(see also Lemma B.1 (iii)). Therefore, with the original labelling,

$$\sum_{k=1}^{n} \{F(b_k)-F(a_k)\} \leqslant F(b)-F(a) = \mu(A).$$

This inequality holds for every positive integer n, and so $\sum_1^\infty \mu(A_n)$ is convergent, and

$$\sum_1^\infty \mu(A_n) \leqslant \mu(A). \tag{5}$$

To obtain the reverse inequality, we proceed as follows. The interval A is made a little smaller and closed, giving, say, B, and the intervals $A_1, A_2, \ldots$ are made a little larger and open, giving, say, $B_1, B_2, \ldots$. Since $B \subseteq \bigcup_1^\infty B_n$, an application of the Heine–Borel theorem shows that, for some positive integer N, $B \subseteq \bigcup_1^N B_n$. By deleting the left-hand end-point of B and appending the right-hand end-points of the B_n's we obtain sets $C, C_1, \ldots, C_N$ of $\mathscr{I}$ for which the corresponding inclusion still holds, and so

$$\mu(C) \leqslant \sum_1^N \mu(C_n).$$

Now B and the B_n's may be chosen so that $\mu(C)$ is nearly equal to $\mu(A)$ and, for each n, $\mu(C_n)$ is nearly equal to $\mu(A_n)$. It follows that

$$\sum_1^N \mu(A_n) > \mu(A) - \text{(a small positive quantity)},$$

and the desired reversed inequality follows.

Now for the details. Let ε be any positive number. Then, by (ii), there exists a positive number $\delta \in (0, b-a)$ such that

$$F(a+\delta)-F(a) < \varepsilon,$$

and positive numbers $\delta_1, \delta_2, \ldots$ such that

$$F(b_n+\delta_n)-F(b_n) < \varepsilon 2^{-n} \qquad (n = 1, 2, \ldots).$$

Let

$$B = [a+\delta, b], \quad C = (a+\delta, b],$$

$$B_n = (a_n, b_n+\delta_n) \qquad (n = 1, 2, \ldots),$$

and

$$C_n = (a_n, b_n+\delta_n] \qquad (n = 1, 2, \ldots).$$

Then

$$B \subseteq A = \bigcup_1^\infty A_n \subseteq \bigcup_1^\infty B_n.$$

Therefore, by the Heine–Borel theorem, there exists a positive integer N such that

$$B \subseteq \bigcup_1^N B_n,$$

and so

$$C \subseteq \bigcup_1^N C_n,$$

i.e.

$$(a+\delta, b] \subseteq \bigcup_1^N (a_n, b_n+\delta_n]. \tag{6}$$

Now let $x_0, x_1, \ldots, x_k$, say, denote the distinct numbers of the set

$$a+\delta,\ b,\ a_1,\ b_1+\delta_1,\ a_2,\ b_2+\delta_2,\ \ldots,\ a_N,\ b_N+\delta_N,$$

where $x_0 < x_1 < \ldots < x_k$. Then

$$\begin{aligned} F(b)-F(a+\delta) &\leqslant F(x_k)-F(x_0) \quad (\text{because } x_0 \leqslant a+\delta < b \leqslant x_k) \\ &= \sum_1^k \{F(x_i)-F(x_{i-1})\}. \end{aligned}$$

By (6), each of the intervals $(x_{i-1}, x_i]$ is contained in at least one of the intervals $C_1, C_2, \ldots, C_N$; choose one of these C_n's which contains $(x_{i-1}, x_i]$, and regard $(x_{i-1}, x_i]$ as being assigned to that particular C_n. Then, for $n = 1, 2, \ldots, N$,

$$\Sigma_n\{F(x_i)-F(x_{i-1})\} \leqslant F(b_n+\delta_n)-F(a_n),$$

where Σ_n denotes summation over those values of i for which $(x_{i-1}, x_i]$ is assigned to C_n (readers requiring a formal proof of this inequality are referred back to the argument used in (4)). Therefore

$$\begin{aligned} \sum_{i=1}^k \{F(x_i)-F(x_{i-1})\} &= \sum_{n=1}^N \Sigma_n\{F(x_i)-F(x_{i-1})\} \\ &\leqslant \sum_{n=1}^N \{F(b_n+\delta_n)-F(a_n)\}, \end{aligned}$$

and so

$$F(b)-F(a+\delta) \leqslant \sum_{n=1}^N \{F(b_n+\delta_n)-F(a_n)\},$$

i.e.

$$\mu(C) \leqslant \sum_{n=1}^N \mu(C_n)$$

(for an alternative proof of this inequality see Lemma B.1 (vi)). Therefore

$$F(b)-F(a)-\varepsilon < \sum_1^N \{F(b_n)-F(a_n)+\varepsilon 2^{-n}\},$$

and so

$$\mu(A) < \sum_1^N \mu(A_n)+\varepsilon(2-2^{-N}).$$

Therefore

$$\mu(A) < \sum_1^\infty \mu(A_n)+2\varepsilon.$$

This holds for all $\varepsilon > 0$, and so

$$\mu(A) \leqslant \sum_1^\infty \mu(A_n). \tag{7}$$

Therefore, by (5) and (7),

$$\mu(A) = \sum_1^{\infty} \mu(A_n),$$

and the proof is complete.

Corollary *Suppose that $F : R \to [0, 1]$ has the properties* (i), (ii) *and* (iii) *of the theorem. Then there exists a random variable X (on some suitably chosen probability space) having the distribution function F.*

Proof. Consider the probability space $(R, \mathscr{B}, P)$, where P is the probability measure on $\mathscr{B}$ for which (3) holds, and let $X : R \to R$ be defined by

$$X(x) = x \quad (x \in R)$$

(i.e. X is the identity function on R). Then it follows from (3) that the distribution function of X is F.

Exercises

1. (i) Suppose that $X : \Omega \to R$ is a random variable, and that

$$P_X(B) = P\{X^{-1}(B)\} \quad (B \in \mathscr{B}).$$

Prove that $(R, \mathscr{B}, P_X)$ is a probability space.

Hint. See Theorem 1.3 (i) and (iii).

(ii) Suppose that $\mathbf{X} = (X_1, X_2, \ldots, X_n)$ is an n-dimensional random variable, i.e. that $X_1, X_2, \ldots, X_n$ are random variables (see Exercise 3.10), and that

$$P_{\mathbf{X}}(B) = P\{\mathbf{X}^{-1}(B)\} \quad (B \in \mathscr{B}^n).$$

Prove that $(R^n, \mathscr{B}^n, P_{\mathbf{X}})$ is a probability space.

2. Suppose that the random variable X has the distribution function F. Prove that, for any real number c,

$$P(X = c) = F(c) - F(c-),$$

where $F(c-) = \lim_{x \to c-} F(x)$.

Hint. $\{X = c\} = \bigcap_{n=1}^{\infty} \left\{c - \frac{1}{n} < X \leqslant c\right\}$.

Deduce that F is continuous at c if and only if $P(X = c) = 0$.

Deduce also that $P(X \in I)$ is

$$\begin{aligned} &F(b-)-F(a) &&\text{if} \quad I = (a, b), \\ &F(b)-F(a-) &&\text{if} \quad I = [a, b], \end{aligned}$$

and

$$F(b-)-F(a-) \quad \text{if} \quad I = [a, b).$$

3. Suppose that the random variable X has the distribution function F.

(i) Prove that X^+, X^- have respectively the distribution functions F_1, F_2 defined by

$$\begin{aligned} &F_1(x) = F_2(x) = 0 &&\text{if} \quad x < 0, \\ &F_1(x) = F(x) &&\text{if} \quad x \geqslant 0, \end{aligned}$$

and

$$F_2(x) = 1-P(X < -x) = 1-F\{(-x)-\} \quad \text{if} \quad x \geqslant 0.$$

(ii) What are the distribution functions of $-X$ and $|X|$?

(iii) Prove that the distribution of X is *symmetric about* 0, i.e. that X and $-X$ have the same distribution, if and only if

$$F(-x) = 1-F(x-) \quad (x \in R).$$

4. The distribution functions F and G are said to be *of the same type* if there exist real numbers a and b such that $a > 0$ and

$$F(ax+b) = G(x) \quad (x \in R).$$

Prove that the relation "is of the same type" is an equivalence relation on the class of all distribution functions.

5. Suppose that X and Y are identically distributed random variables, and that $f: R \to R$ is Borel measurable. Prove that $f(X)$ and $f(Y)$ are identically distributed random variables.

6. Suppose that X and Y are identically distributed random variables, and that $A = X^{-1}(C)$, $B = Y^{-1}(C)$ for some given Borel set C. Prove that XI_A and YI_B are identically distributed random variables.

Hint. Take $f(x) = xI_C(x) \quad (x \in R)$ in Exercise 5.

7. Let F be a distribution function, and let

$$F_h(x) = \frac{1}{2h} \int_{x-h}^{x+h} F(t)\, dt \quad (x \in R),$$

where h is a positive constant. Prove that F_h is a distribution function which is everywhere continuous.

8. Show that the distribution function F of a random variable X takes only the two values 0 and 1 if and only if there exists a real number c such that $P(X = c) = 1$.

Hint. ("only if") Consider $\inf\{x : F(x) = 1\}$.

9. Let X be a random variable with distribution function F. By considering $\sup\{x : F(x) < \frac{1}{2}\}$, show that X has at least one median. (A *median* of X is any real number m for which $P(X \leqslant m)$ and $P(X \geqslant m)$ are both $\geqslant \frac{1}{2}$, or, equivalently, for which $F(m-) \leqslant \frac{1}{2} \leqslant F(m)$.) Show also that the median you have just obtained is the smallest median of X.

By considering similarly $\inf\{x : F(x) > \frac{1}{2}\}$, show that X has a greatest median, and deduce that the set of all medians of X consists either of a single point or a finite closed interval.

10. Let F be the distribution function of a random variable X, and let $Y = F(X)$. Prove that if F is continuous on R then Y has the distribution function G determined by $G(y) = y\ (0 < y < 1)$.

Note. A random variable with this distribution function G is said to be uniformly distributed on (0, 1) (see also Exercise 5.23 (i)).

11. Suppose that P_1 and P_2 are probability measures on $\mathscr{B}^n$ for which

$$P_1(A) = P_2(A) \quad (A \in \mathscr{I}^n).$$

Prove that $P_1 = P_2$ on $\mathscr{B}^n$.

12. Suppose that $\mathbf{X} = (X_1, X_2, \ldots, X_n) : \Omega \to R^n$ is an n-dimensional random variable (see Exercise 1(ii)). The distribution function $F : R^n \to [0, 1]$ of $\mathbf{X}$ is defined by

$$F(x_1, x_2, \ldots, x_n) = P\{X_i \leqslant x_i\ (i = 1, 2, \ldots, n)\}\ (x_1, x_2, \ldots, x_n \in R).$$

Prove the following results (cf. Theorem 1):

(i) If $\Delta_1, \Delta_2, \ldots, \Delta_n$ are the difference operators defined by

$$\begin{aligned}\Delta_i(h) f(x_1, x_2, \ldots, x_n) = &f(x_1, \ldots, x_{i-1}, x_i+h, x_{i+1}, \ldots, x_n)\\ &-f(x_1, \ldots, x_{i-1}, x_i, x_{i+1}, \ldots, x_n),\end{aligned}$$

then

$$\Delta_1(h_1)\,\Delta_2(h_2) \ldots \Delta_n(h_n)\, F(x_1, x_2, \ldots, x_n) \geqslant 0 \qquad \text{(a)}$$

for any real numbers $x_1, x_2, \ldots, x_n$ and positive numbers $h_1, h_2, \ldots, h_n$.

Hint. Verify that the left-hand side of (a) is

$$P\{X_i \in (x_i, x_i+h_i]\ (i = 1, 2, \ldots, n)\}.$$

(ii) F is non-decreasing in each variable separately, i.e.

$$\Delta_i(h)\, F(x_1, x_2, \ldots, x_n) \geqslant 0$$
$$(x_1, x_2, \ldots, x_n \in R;\ i = 1, 2, \ldots, n;\ h > 0).$$

(iii) F is everywhere continuous on the right in each variable separately, i.e.

$$\Delta_i(h)\, F(x_1, x_2, \ldots, x_n) \to 0 \quad \text{as} \quad h \to 0+$$
$$(x_1, x_2, \ldots, x_n \in R;\ i = 1, 2, \ldots, n).$$

(iv) $F(x_1, x_2, \ldots, x_n) \to 0$ if *any one* of $x_1, x_2, \ldots, x_n \to -\infty$, or if $\min(x_1, x_2, \ldots, x_n) \to -\infty$, and $F(x_1, x_2, \ldots, x_n) \to 1$ if *all* of $x_1, x_2, \ldots, x_n \to \infty$, i.e. if $\min(x_1, x_2, \ldots, x_n) \to \infty$.

Note. The analogue of Theorem 3 holds for n-dimensional distribution functions. More precisely, if $F : R^n \to [0, 1]$ satisfies (i), (iii) and (iv) above, there exists one and only one probability measure P on $\mathscr{B}^n$ such that

$$P\{(-\infty, x_1] \times (-\infty, x_2] \times \ldots \times (-\infty, x_n]\}$$
$$= F(x_1, x_2, \ldots, x_n) \quad (x_1, x_2, \ldots, x_n \in R).$$

It will be an instructive exercise for the reader to justify this assertion for, say, $n = 2$ by suitably modifying the proof of Theorem 3.

5 Expected values

Let X be the number shown when a die is thrown. Then the expected value (mean value, expectation) $E(X)$ of X is the average of the numbers shown in a long series of throws of the die. If the die is fair, i.e. if $P(X = i) = \frac{1}{6}$ for $i = 1, 2, \ldots, 6$, then

$$E(X) = 1\times\tfrac{1}{6}+2\times\tfrac{1}{6}+ \ldots +6\times\tfrac{1}{6} = 3\tfrac{1}{2}.$$

More generally, if the probability of X taking each value i is not assumed to be the same then

$$E(X) = \sum_{i=1}^{6} ip_i,$$

where $p_i = P(X = i)$ $(i = 1, 2, \ldots, 6)$.

This definition of $E(X)$ extends at once to any random variable which takes only a finite number of distinct values (i.e. a simple random variable). It is the aim of this chapter to extend the definition to random variables of more general type, and to obtain rules for operating with expected values.

The definition takes place in three stages. Firstly, $E(X)$ is defined for simple (and, in fact, non-negative) random variables X. Secondly, if X is any non-negative random variable, $E(X)$ is defined as

$$\lim_{n\to\infty} E(X_n),$$

where $\{X_n\}$ is a monotone increasing sequence of simple non-negative random variables with limit X (the existence of at least one such sequence was established in Theorem 3.7); thus $E(X)$ is approximated "from below". Thirdly, if X is a random variable which can take values of either sign, $E(X)$ is defined (under certain circumstances) as $E(X^+)-E(X^-)$ (note that X^+ and X^- are non-negative random variables, and that $X = X^+ - X^-$). The above programme is not merely a convenient way of defining $E(X)$ when it exists, it is also, as the reader will see in due course, a convenient way of proving theorems about expected values. These are proved first for simple non-negative random variables, when they are often trivial, and then extended successively to general non-negative random variables and to random variables which can take values of either sign.

A final extension of the definition of $E(X)$ will be made to "complex-valued" random variables; these will be needed later for the treatment of characteristic functions.

We shall assume, unless stated otherwise (as, for example, in Theorem 10), that all random variables are defined on the probability space $(\Omega, \mathscr{F}, P)$.

(I) Suppose that the random variable X is simple and non-negative. Then X is of the form

$$a_1I_{A_1}+a_2I_{A_2}+ \dots +a_mI_{A_m}, \tag{1}$$

where $$a_1, a_2, \dots, a_m \geqslant 0$$

and $$\{A_1, A_2, \dots, A_m\} \quad \text{is an } \mathscr{F}\text{-partition of } \Omega.$$

We define $E(X)$ as follows.

Definition 1. $E(X) = \sum_{i=1}^{m} a_iP(A_i)$.

Clearly $E(X)$ is non-negative and finite: $0 \leqslant E(X) < \infty$.

If we require the a's to be distinct, then the form (1) is unique and $E(X)$ is uniquely determined by the above definition. However, it is not always convenient to require this, and so before going any further it will be shown that $E(X)$ does not depend on the particular expression (1).

Suppose that X is also equal to

$$b_1I_{B_1}+b_2I_{B_2}+ \dots +b_nI_{B_n},$$

where $$b_1, b_2, \dots, b_n \geqslant 0$$

and $$\{B_1, B_2, \dots, B_n\} \quad \text{is an } \mathscr{F}\text{-partition of } \Omega.$$

Now, for $j = 1, 2, \dots, n$,

$$B_j = \bigcup_{i=1}^{m} A_i \cap B_j,$$

because $\{A_1, A_2, \dots, A_m\}$ is a partition of Ω, and so

$$P(B_j) = \sum_{i=1}^{m} P(A_i \cap B_j),$$

because the sets $A_i \cap B_j$ $(i = 1, 2, \dots, m)$ are disjoint. Similarly

$$P(A_i) = \sum_{j=1}^{n} P(A_i \cap B_j) \qquad (i = 1, 2, \dots, m).$$

Therefore

$$\sum_{i=1}^{m} a_i P(A_i) = \sum_{i=1}^{m} \sum_{j=1}^{n} a_i P(A_i \cap B_j)$$

and

$$\sum_{j=1}^{n} b_j P(B_j) = \sum_{i=1}^{m} \sum_{j=1}^{n} b_j P(A_i \cap B_j).$$

Now, for all i and j,

$$a_i P(A_i \cap B_j) = b_j P(A_i \cap B_j);$$

for this is immediate if $A_i \cap B_j = \emptyset$, and if $A_i \cap B_j \neq \emptyset$ then $a_i = b_j = X(\omega)$, where ω is any point of $A_i \cap B_j$. Therefore

$$\sum_{i=1}^{m} a_i P(A_i) = \sum_{j=1}^{n} b_j P(B_j),$$

and so the value of $E(X)$ does not depend on the particular expression (1) used in its definition.

The reader should note that

$$E(I_A) = P(A)$$

for all $A \in \mathscr{F}$.

Lemma 1. *Suppose that the random variables X and Y are simple and non-negative, and that c is any non-negative real number. Then*

(i) $E(cX) = cE(X)$;
(ii) $E(X+Y) = E(X)+E(Y)$;
(iii) *if* $X \geqslant Y$ *on* Ω, $E(X) \geqslant E(Y)$.

Proof. The proof of (i) is left to the reader.

To prove (ii) suppose that

$$X = \sum_{i=1}^{m} a_i I_{A_i} \quad \text{and} \quad Y = \sum_{j=1}^{n} b_j I_{B_j},$$

where the a's, b's, A's and B's satisfy the appropriate conditions. Then

$$X = \sum_{i=1}^{m} \sum_{j=1}^{n} a_i I_{A_i \cap B_j} \quad \text{and} \quad Y = \sum_{i=1}^{m} \sum_{j=1}^{n} b_j I_{A_i \cap B_j},$$

and the result follows immediately from Definition 1. (The reader will note that we make use here of the fact that the expected value of a simple non-negative random variable does not depend on the particular form in which it is expressed.)

To prove (iii) we can either use the expressions for X and Y which we have just obtained, and note that $a_i \geqslant b_j$ whenever $A_i \cap B_j \neq \emptyset$, or note that $X-Y$ is a simple non-negative random variable and so, by (ii),

$$E(X) = E(Y)+E(X-Y) \geqslant E(Y).$$

The details are left to the reader.

Lemma 2. *Suppose that* $X_1, X_2, \ldots$ *and* Y *are simple random variables satisfying*

$$0 \leqslant X_1 \leqslant X_2 \leqslant \ldots \text{ on } \Omega \tag{2}$$

and

$$0 \leqslant Y \leqslant \lim_{n\to\infty} X_n \text{ on } \Omega. \tag{3}$$

Then

$$\lim_{n\to\infty} E(X_n) \geqslant E(Y).$$

Notes. (a) The statement (2) is equivalent to the statement that

$$0 \leqslant X_1(\omega) \leqslant X_2(\omega) \leqslant \ldots$$

for all $\omega \in \Omega$. It follows that, for all $\omega \in \Omega$, $X_n(\omega)$ tends to a limit, which may be $+\infty$, as $n \to \infty$.

(b) It follows from (2) and Lemma 1 (iii) that

$$E(X_1) \leqslant E(X_2) \leqslant \ldots,$$

and so $E(X_n)$ tends to a limit as $n \to \infty$ (the limit may be $+\infty$, in which case the conclusion of the theorem is trivial).

Proof. Suppose that $\varepsilon > 0$, and let

$$A_n = \{X_n+\varepsilon > Y\} \qquad (n = 1, 2, \ldots).$$

Then, for every positive integer n,

$$X_n I_{A_n}+\varepsilon I_{A_n} \geqslant Y I_{A_n} \text{ on } \Omega$$

(for $X_n+\varepsilon > Y$ on A_n, and I_{A_n} vanishes on A_n^c), and so, by Lemma 1 (ii) and (iii),

$$E(X_n I_{A_n})+E(\varepsilon I_{A_n}) \geqslant E(Y I_{A_n}).$$

Now $X_n I_{A_n} \leqslant X_n$, and so $E(X_n I_{A_n}) \leqslant E(X_n)$ (by Lemma 1 (iii)), and $E(\varepsilon I_{A_n}) = \varepsilon P(A_n) \leqslant \varepsilon$. Therefore

$$E(X_n) \geqslant E(Y I_{A_n})-\varepsilon \qquad (n = 1, 2, \ldots). \tag{4}$$

Now suppose that

$$Y = \sum_{j=1}^{k} c_j I_{B_j},$$

where $c_1, c_2, \ldots, c_k \geqslant 0$ and $\{B_1, B_2, \ldots, B_k\}$ is an $\mathscr{F}$-partition of Ω. Then

$$YI_{A_n} = \sum_{j=1}^{k} c_j I_{B_j} I_{A_n} = \sum_{j=1}^{k} c_j I_{B_j \cap A_n},$$

and so

$$E(YI_{A_n}) = \sum_{j=1}^{k} c_j P(B_j \cap A_n).$$

It follows from (2) and (3) and the definition of the A's that

$$A_1 \subseteq A_2 \subseteq \ldots \quad \text{and} \quad \bigcup_{n=1}^{\infty} A_n = \Omega,$$

and so, for $j = 1, 2, \ldots, k$,

$$B_j \cap A_1 \subseteq B_j \cap A_2 \subseteq \ldots \quad \text{and} \quad \bigcup_{n=1}^{\infty} B_j \cap A_n = B_j.$$

Therefore

$$P(B_j \cap A_n) \to P(B_j) \quad \text{as} \quad n \to \infty \qquad (j = 1, 2, \ldots, k)$$

(by Theorem 2.2 (i)), and so

$$E(YI_{A_n}) \to \sum_{j=1}^{k} c_j P(B_j) = E(Y) \quad \text{as} \quad n \to \infty.$$

It now follows from (4) that

$$\lim_{n \to \infty} E(X_n) \geqslant E(Y) - \varepsilon.$$

This holds for all $\varepsilon > 0$, and so

$$\lim_{n \to \infty} E(X_n) \geqslant E(Y).$$

(II) Now suppose that the random variable X is non-negative. By Theorem 3.7, there exist simple random variables $X_1, X_2, \ldots$ satisfying

$$0 \leqslant X_1 \leqslant X_2 \leqslant \ldots \text{ on } \Omega \tag{5}$$

and

$$X_n \to X \text{ on } \Omega \quad \text{as} \quad n \to \infty. \tag{6}$$

By Lemma 1(iii),

$$0 \leqslant E(X_1) \leqslant E(X_2) \leqslant \ldots,$$

and so $E(X_n)$ tends to a limit, which may be $+\infty$, as $n \to \infty$.

Definition 2. $E(X) = \lim_{n \to \infty} E(X_n)$.

($E(X_1), E(X_2), \ldots$ are to be interpreted in the sense of Definition 1.)

Again $E(X)$ is non-negative, but now it may be infinite: $0 \leqslant E(X) \leqslant \infty$.

It will be shown firstly that every sequence $\{X_n\}$ of simple non-negative random variables satisfying (5) and (6) leads to the same value of $E(X)$, and secondly that Definitions 1 and 2 lead to the same value of $E(X)$ when X is a simple non-negative random variable.

To prove the first of these statements, let $\{X_n'\}$ be any sequence of simple non-negative random variables satisfying

$$0 \leqslant X_1' \leqslant X_2' \leqslant \ldots \text{ on } \Omega$$

and

$$X_n' \to X \text{ on } \Omega \quad \text{as} \quad n \to \infty.$$

Then by Lemma 2, with $Y = X_m'$,

$$\lim_{n \to \infty} E(X_n) \geqslant E(X_m') \qquad (m = 1, 2, \ldots).$$

On letting m tend to infinity it follows that

$$\lim_{n \to \infty} E(X_n) \geqslant \lim_{m \to \infty} E(X_m').$$

Similarly

$$\lim_{m \to \infty} E(X_m') \geqslant \lim_{n \to \infty} E(X_n),$$

and so

$$\lim_{m \to \infty} E(X_m') = \lim_{n \to \infty} E(X_n).$$

Thus the same value of $E(X)$ is obtained, whether it is defined by means of the sequence $\{X_n\}$ or by means of the sequence $\{X_n'\}$.

Now let X be a simple non-negative random variable. Then we can take $X_1 = X_2 = \ldots = X$ in Definition 2, and so Definitions 1 and 2 both lead to the same value of $E(X)$.

Lemma 3. *Suppose that the random variables X and Y are non-negative, and that c is a non-negative real number. Then*

(i) $E(cX) = cE(X)$

where $cE(X)$ is defined to be 0 *if* $c = 0$ *and* $E(X) = \infty$;

(ii) $E(X+Y) = E(X)+E(Y)$;

(iii) *if* $X \geqslant Y$ *on* Ω, $E(X) \geqslant E(Y)$.

Proof. Choose simple random variables $X_1, X_2, \ldots$ such that (5) and (6) hold, and simple random variables $Y_1, Y_2, \ldots$ such that

$$0 \leqslant Y_1 \leqslant Y_2 \leqslant \ldots \text{ on } \Omega$$

and

$$Y_n \to Y \text{ on } \Omega \quad \text{as} \quad n \to \infty.$$

(i) We note that $cX_1, cX_2, \ldots$ are simple random variables,

$$0 \leqslant cX_1 \leqslant cX_2 \leqslant \ldots \text{ on } \Omega,$$

and

$$cX_n \to cX \text{ on } \Omega \quad \text{as} \quad n \to \infty.$$

Therefore

$$\begin{aligned} E(cX) &= \lim_{n\to\infty} E(cX_n) \quad &&\text{(by Definition 2)} \\ &= \lim_{n\to\infty} cE(X_n) &&\text{(by Lemma 1 (i))} \\ &= cE(X) &&\text{(by Definition 2).} \end{aligned}$$

(ii) The proof of (ii) is similar, $E(X+Y)$ being of defined by means of the sequence $\{X_n+Y_n\}$.

(iii) As in the proof of Lemma 1 (iii) we note that $X-Y$ is a non-negative random variable, and so, by (ii),

$$E(X) = E(Y)+E(X-Y) \geqslant E(Y).$$

(III) Now suppose that the random variable X can take values of either sign. Then

$$X^+ = \tfrac{1}{2}(|X|+X), \quad X^- = \tfrac{1}{2}(|X|-X)$$

are non-negative random variables (by Theorem 3.3 (ii)).

Definition 3. $E(X)$ is said to *exist* if $E(X^+)$ and $E(X^-)$ are not both infinite, and then

$$E(X) = E(X^+)-E(X^-).$$

($E(X^+)$ and $E(X^-)$ are to be interpreted in the sense of Definition 2.)

Notes. (a) $E(X)$ is not defined if $E(X^+) = E(X^-) = \infty$.

(b) $E(X) = \infty$ if and only if $E(X^+) = \infty$ and $E(X^-)$ is finite,
$E(X) = -\infty$ if and only if $E(X^+)$ is finite and $E(X^-) = \infty$,
and $E(X)$ is *finite* if and only if $E(X^+)$ and $E(X^-)$ are both finite.

(c) $E(X)$ certainly exists if $X \geqslant 0$ on Ω (when $X^- = 0$ on Ω) or if $X \leqslant 0$ on Ω (when $X^+ = 0$ on Ω).

This time there is no need to prove that $E(X)$ is defined uniquely. However, we must show that Definitions 2 and 3 give the same value for $E(X)$ when X is a non-negative random variable. This is almost immediate, for when X is non-negative, $X^+ = X$ and $X^- = 0$ (and so $E(X^-) = 0$).

As an elementary exercise on Definition 3, the reader should prove that if $X = c$ (a constant) on Ω then $E(X) = c$.

Those readers with some previous knowledge of the general theory of integration will recognise that $E(X)$ is the integral of the function X over Ω with respect to the measure P, and as such is denoted by

$$\int_\Omega X\,dP \quad \text{or} \quad \int_\Omega X(\omega)\,dP.$$

This notation will occasionally be found useful when more than one probability measure P is involved (for example, in Theorem 10).

For the remainder of this book, any reference to $E(X)$ (e.g. "$E(X)$ exists", "$E(X)$ is finite") will automatically imply that X is a random variable.

Theorem 1 *Suppose that $E(X)$ is finite and that c is a real number. Then $E(cX)$ is finite and*

$$E(cX) = cE(X).$$

Proof. This follows from Lemma 3 (i) on noting that

$$(cX)^+ = cX^+, \quad (cX)^- = cX^- \quad \text{if} \quad c \geqslant 0$$

and

$$(cX)^+ = (-c)X^-, \quad (cX)^- = (-c)X^+ \quad \text{if} \quad c < 0.$$

Theorem 2 *Suppose that $E(X)$ and $E(Y)$ are finite. Then $E(X+Y)$ is finite and*

$$E(X+Y) = E(X)+E(Y). \tag{7}$$

Proof. Since
$$\begin{aligned}(X+Y)^+ &= \tfrac{1}{2}\{|X+Y|+(X+Y)\}\\ &\leqslant \tfrac{1}{2}(|X|+|Y|+X+Y) = X^+ + Y^+,\end{aligned}$$

it follows that

$$\begin{aligned}E\{(X+Y)^+\} &\leqslant E(X^+ + Y^+) && \text{(by Lemma 3 (iii))}\\ &= E(X^+)+E(Y^+) && \text{(by Lemma 3 (ii)).}\end{aligned}$$

Now $E(X^+)$ and $E(Y^+)$ are finite (because $E(X)$ and $E(Y)$ are), and so $E\{(X+Y)^+\}$ is finite. Similarly $E\{(X+Y)^-\}$ is finite, and so $E(X+Y)$ is finite.

To prove (7) we note that

$$(X+Y)^+ + X^- + Y^- = (X+Y)^- + X^+ + Y^+$$

(since each side is equal to $\frac{1}{2}(|X+Y|+|X|+|Y|)$), and so, by Lemma 3 (ii),

$$E\{(X+Y)^+\}+E(X^-)+E(Y^-) = E\{(X+Y)^-\}+E(X^+)+E(Y^+). \quad (8)$$

Since all the terms in this equation are finite

$$E\{(X+Y)^+\}-E\{(X+Y)^-\} = \{E(X^+)-E(X^-)\}+\{E(Y^+)-E(Y^-)\},$$

and (7) follows.

Note. It does not follow from (8) that $E\{(X+Y)^+\}$ and $E\{(X+Y)^-\}$ are finite, since (8) is consistent with both these expected values being infinite. It is thus necessary to show specifically that $E(X+Y)$ is finite.

Theorem 3 *Suppose that* $E(X)$ *and* $E(Y)$ *exist, and that* $X \geqslant Y$ *on* Ω. *Then*

$$E(X) \geqslant E(Y).$$

Proof. By considering the various possibilities for the signs of X and Y at any point ω of Ω (e.g. $X(\omega) \geqslant 0$, $Y(\omega) < 0$), it is easy to see that

$$X^+ \geqslant Y^+ \quad \text{and} \quad X^- \leqslant Y^- \text{ on } \Omega.$$

Therefore, by Lemma 3 (iii),

$$E(X^+) \geqslant E(Y^+) \quad \text{and} \quad E(X^-) \leqslant E(Y^-). \quad (9)$$

Now if $E(X^+) = \infty$ (and so $E(X) = \infty$), or if $E(Y^-) = \infty$ (and so $E(Y) = -\infty$), the result is trivial. Otherwise, by (9), $E(X^+)$, $E(Y^+)$, $E(Y^-)$ and $E(X^-)$ are all finite, and, again by (9),

$$E(X^+)-E(X^-) \geqslant E(Y^+)-E(Y^-),$$

which gives the desired result.

Theorem 4 (i) *Suppose that* X *is a random variable. Then* $E(X)$ *is finite if and only if* $E(|X|)$ *is finite.*

(ii) *Suppose that* $E(X)$ *is finite. Then*

$$|E(X)| \leqslant E(|X|).$$

(iii) *Suppose that* X *is a random variable,* $E(Y)$ *is finite, and* $|X| \leqslant Y$ *on* Ω. *Then* $E(X)$ *is finite.*

Note. In (i) and (iii) we assume explicitly that X is a random variable, i.e. that X is measurable. This is clearly necessary in (iii); it is necessary in (i) since the measurability of $|X|$ (which is implied by the statement that $E(|X|)$ is finite) does not entail that of X.

Proof. (i) Suppose that $E(X)$ is finite. Then $E(X^+)$ and $E(X^-)$ are finite (by Definition 3). Now $|X| = X^+ + X^-$, and so, by Theorem 2, $E(|X|)$ is finite.

Conversely, suppose that $E(|X|)$ is finite. Since $0 \leqslant X^+, X^- \leqslant |X|$ on Ω, it follows from Theorem 3 (or Lemma 3 (iii)) that $E(X^+)$ and $E(X^-)$ are finite, and so, by Definition 3, that $E(X)$ is finite.

(ii)
$$\begin{aligned} |E(X)| &= |E(X^+) - E(X^-)| && \text{(by Definition 3)} \\ &\leqslant E(X^+) + E(X^-) && \text{(because } E(X^+) \geqslant 0 \text{ and } E(X^-) \geqslant 0) \\ &= E(|X|) && \text{(by Theorem 2).} \end{aligned}$$

(iii) Since $0 \leqslant |X| \leqslant Y$ on Ω, it follows from Theorem 3 that $E(|X|)$ is finite, and so, by (i), that $E(X)$ is finite.

Corollary *Suppose that the random variable X is bounded. Then $E(X)$ is finite*

Proof. If, say, $|X| \leqslant c$ on Ω, then (iii) can be applied with the random variable Y defined by $Y(\omega) = c \ (\omega \in \Omega)$.

Theorem 5 (*the monotone convergence theorem*) *Suppose that $X_1, X_2, \ldots$ are random variables such that*

$$0 \leqslant X_1 \leqslant X_2 \leqslant \ldots \text{ on } \Omega$$

and $\quad X_n(\omega)$ *tends to a limit* $X(\omega)$ *as* $n \to \infty \quad (\omega \in \Omega)$.

Then
$$E(X_n) \to E(X) \quad \text{as} \quad n \to \infty.$$

Notes. (a) X is a random variable (by Theorem 3.6 (iii)).

(b) Since $X_1, X_2, \ldots$ and X are non-negative on Ω, all the expected values exist.

(c) In the special case when $X_1, X_2, \ldots$ are all simple, the result is a restatement of Definition 2 of $E(X)$. In the general case the proof consists essentially of replacing $X_1, X_2, \ldots$ by simple random variables $Y_1, Y_2, \ldots$ which satisfy $0 \leqslant Y_1 \leqslant Y_2 \leqslant \ldots$ on Ω, $Y_n \to X$ on Ω as $n \to \infty$, and $\lim_{n\to\infty} E(Y_n) = \lim_{n\to\infty} E(X_n)$.

Proof. Since $0 \leqslant X_1 \leqslant X_2 \leqslant \ldots \leqslant X$ on Ω, it follows from Theorem 3 that

$$0 \leqslant E(X_1) \leqslant E(X_2) \leqslant \ldots \leqslant E(X).$$

Therefore $E(X_n)$ tends to a limit (which may be $+\infty$) as $n \to \infty$, and

$$\lim_{n\to\infty} E(X_n) \leqslant E(X).$$

It remains to prove that

$$\lim_{n\to\infty} E(X_n) \geqslant E(X).$$

For each positive integer m there exist simple random variables X_{m1}, X_{m2}, ... for which

$$0 \leqslant X_{m1} \leqslant X_{m2} \leqslant \ldots \text{ on } \Omega$$

and

$$X_{mn} \to X_m \text{ on } \Omega \quad \text{as} \quad n \to \infty.$$

For $n = 1, 2, \ldots$ let

$$Y_n = \max(X_{1n}, X_{2n}, \ldots, X_{nn}).$$

Then it is easily verified that $Y_1, Y_2, \ldots$ are simple random variables for which

$$0 \leqslant Y_1 \leqslant Y_2 \leqslant \ldots \text{ on } \Omega, \tag{10}$$

and that

$$X_{mn} \leqslant Y_n \leqslant X_n \quad \text{on} \quad \Omega \qquad (n = 1, 2, \ldots;\ m = 1, 2, \ldots, n). \tag{11}$$

On letting n tend to infinity it follows that

$$X_m \leqslant \lim_{n\to\infty} Y_n \leqslant X \quad \text{on} \quad \Omega$$

for $m = 1, 2, \ldots$ (the existence of $\lim_{n\to\infty} Y_n$ follows from (10)). Now, on letting m tend to infinity, it follows that

$$\lim_{n\to\infty} Y_n = X \quad \text{on} \quad \Omega.$$

It follows from Definition 2 that

$$\lim_{n\to\infty} E(Y_n) = E(X).$$

But $X_n \geqslant Y_n$ on Ω (by (11)), and so $E(X_n) \geqslant E(Y_n)$ $(n = 1, 2, \ldots)$. Consequently $\lim_{n\to\infty} E(X_n) \geqslant E(X)$, and the proof is complete.

Theorem 6 (*the dominated convergence theorem*) *Suppose that*

(i) $X_1, X_2, \ldots$ *are random variables*;

(ii) *there exists a random variable Y such that E(Y) is finite and*

$$|X_n| \leqslant Y \text{ on } \Omega \qquad (n = 1, 2, \ldots);$$

(iii) X_n *tends to a limit X on* Ω *as* $n \to \infty$.

Then E(X) is finite and

$$E(X_n) \to E(X) \quad \text{as} \quad n \to \infty.$$

Notes. (a) X is a random variable (by Theorem 3.6 (iii)).

(b) $E(X_n)$ is finite for all n (by Theorem 4 (iii)).

Proof. It follows from (ii) and the definition of X that $|X| \leqslant Y$ on Ω, and so $E(X)$ is finite (by Theorem 4 (iii)).

For $n = 1, 2, \ldots$ let

$$Y_n = |X_n - X|$$

and

$$Z_n = \sup(Y_n, Y_{n+1}, \ldots).$$

Now $0 \leqslant Y_m \leqslant 2Y$ on Ω (all m), and so Z_n is a random variable (by Theorem 3.6 (i)) and

$$0 \leqslant Z_n \leqslant 2Y \quad \text{on} \quad \Omega. \tag{12}$$

Furthermore, by the definition of the Z's,

$$Z_1 \geqslant Z_2 \geqslant \ldots \quad \text{on} \quad \Omega,$$

and so

$$0 \leqslant 2Y - Z_1 \leqslant 2Y - Z_2 \leqslant \ldots \text{ on } \Omega.$$

Now, for each $\omega \in \Omega$, $Y_n(\omega) \to 0$ as $n \to \infty$ (by (iii)), and so $Z_n(\omega) \to 0$ as $n \to \infty$ (the reader should prove this). Therefore

$$2Y - Z_n \to 2Y \text{ on } \Omega \quad \text{as} \quad n \to \infty.$$

We can now apply the monotone convergence theorem to obtain

$$E(2Y - Z_n) \to E(2Y) \quad \text{as} \quad n \to \infty.$$

Since $E(2Y)$ and $E(Z_n)$ are finite (the finiteness of $E(Z_n)$ follows from (12)), and so

$$E(2Y - Z_n) = E(2Y) - E(Z_n),$$

it follows that

$$E(Z_n) \to 0 \quad \text{as} \quad n \to \infty.$$

Therefore

$$\begin{aligned} |E(X_n)-E(X)| &= |E(X_n-X)| \\ &\leqslant E(|X_n-X|) \quad \text{(by Theorem 4 (ii))} \\ &= E(Y_n) \\ &\leqslant E(Z_n) \quad \text{(because } 0 \leqslant Y_n \leqslant Z_n \text{ on } \Omega) \\ &\to 0 \quad \text{as} \quad n \to \infty, \end{aligned}$$

and the proof is complete.

Note. The basic idea underlying the proof of this theorem is that of obtaining a suitable sequence of random variables to which the monotone convergence theorem can be applied. The sequence $\{Y_n\}$ converges to zero on Ω, but not necessarily monotonically. The sequence $\{Z_n\}$ decreases steadily to zero on Ω. Finally the sequence $\{2Y-Z_n\}$, to which Theorem 5 is applicable, is obtained.

Corollary (*the bounded convergence theorem*) *Suppose that*

(i) $X_1, X_2, \ldots$ *are random variables*;
(ii) *there exists a real number c such that*

$$|X_n| \leqslant c \text{ on } \Omega \qquad (n = 1, 2, \ldots);$$

(iii) X_n *tends to a limit* X *on* Ω *as* $n \to \infty$.

Then $E(X)$ *is finite and*

$$E(X_n) \to E(X) \quad \text{as} \quad n \to \infty.$$

Proof. Apply the theorem with $Y(\omega) = c$ $(\omega \in \Omega)$.

In the general theory of integration, the monotone and dominated convergence theorems are the two basic results which give conditions that justify the interchange of $\int$ and $\lim_{n\to\infty}$, i.e. that ensure that

$$\int \lim_{n\to\infty} f_n = \lim_{n\to\infty} \int f_n.$$

(The special case of the Lebesgue integral is, in fact, mentioned in Appendix A.) Theorems 5 and 6 are restatements of these two basic results in random variable language. Their importance lies in their providing conditions which justify the carrying out of various limiting operations "under the expectation sign", for example, the summation of an infinite series (Exercise 16), differentiation (Theorem 8) or integration (Theorem 9).

In the next three theorems we consider random variables which depend on a parameter t, and whose expected values will, therefore, be functions of t. More precisely, we consider functions of two variables of the form $X: I \times \Omega \to R$, where I is an interval on the real line, which have the property that the function $X^{(t)}: \Omega \to R$ defined by

$$X^{(t)}(\omega) = X(t; \omega) \qquad (t \in I; \omega \in \Omega)$$

is a random variable for each fixed $t \in I$. If we suppose that $E(X^{(t)})$ is finite for all $t \in I$, and let $F(t) = E(X^{(t)})$, then $F: I \to R$. The theorems give conditions which ensure that F is continuous on I, or justify the operations of differentiation or integration "under the expectation sign".

In each of the following three theorems we assume that

(A) I is an interval (finite or infinite) of the real line;

(B) the function $X: I \times \Omega \to R$ has the property that the function $X^{(t)}: \Omega \to R$ as defined above is a random variable for each fixed $t \in I$.

Theorem 7 *Suppose that* (A) *and* (B) *above hold, and suppose also that*

(i) *there exists a random variable Y such that $E(Y)$ is finite and*

$$|X^{(t)}| \leqslant Y \text{ on } \Omega \qquad (t \in I);$$

(ii) *for each fixed $\omega \in \Omega$, $X(t; \omega)$ is continuous on I.*
Then $F(t) = E\{X^{(t)}\}$ is continuous on I.

Note. It follows from Theorem 4 (iii) that F is finite on I.

Proof. Let t_0 be any point of I. By a standard theorem on continuity it suffices to prove that

$$F(t_n) \to F(t_0) \quad \text{as} \quad n \to \infty \tag{13}$$

whenever

$$t_1, t_2, \ldots \in I \quad \text{and} \quad t_n \to t_0 \text{ as } n \to \infty. \tag{14}$$

Suppose that (14) holds, and let $X_n = X^{(t_n)}$ $(n = 0, 1, 2, \ldots)$. Then, by (ii), $X_n \to X_0$ on Ω as $n \to \infty$. The dominated convergence theorem now gives $E(X_n) \to E(X_0)$ as $n \to \infty$. Thus (13) holds, and the proof is complete.

In the next theorem, the function X is assumed to be differentiable on the interval I for each fixed $\omega \in \Omega$. The derivative

$$\lim_{h \to 0} \frac{X(t+h; \omega) - X(t; \omega)}{h}$$

will be denoted by $\frac{\partial}{\partial t} X(t; \omega)$. The first t is part of the name of the function

$\left(\frac{\partial}{\partial t}X\right)$; the second t is a point of the interval I. The notation may be considered unfortunate, but is sanctified by usage.

As before, $\frac{\partial}{\partial t}X^{(t)}: \Omega \to R$ is defined by

$$\frac{\partial}{\partial t}X^{(t)}(\omega) = \frac{\partial}{\partial t}X(t;\omega) \quad (t \in I;\ \omega \in \Omega).$$

Theorem 8 *Suppose that* (A) *and* (B) *above hold, and suppose also that*

(i) *I is open;*
(ii) *$F(t) = E\{X^{(t)}\}$ is finite* $(t \in I)$;
(iii) *for each fixed $\omega \in \Omega$, $X(t;\omega)$ has a finite derivative at each point of I;*
(iv) *there exists a random variable Y such that $E(Y)$ is finite and*

$$\left|\frac{\partial}{\partial t}X^{(t)}\right| \leqslant Y \text{ on } \Omega \quad (t \in I).$$

Then $E\left\{\frac{\partial}{\partial t}X^{(t)}\right\}$ is finite on I, and

$$F'(t) = E\left\{\frac{\partial}{\partial t}X^{(t)}\right\} \text{ on } I,$$

i.e.
$$\frac{d}{dt}E\{X^{(t)}\} = E\left\{\frac{\partial}{\partial t}X^{(t)}\right\} \text{ on } I.$$

Note. It will appear in the course of the proof that $\frac{\partial}{\partial t}X^{(t)}$ is, in fact, a random variable for each fixed $t \in I$.

Proof. Let t_0 be any point of I, and let

$$X_0(\omega) = \frac{\partial}{\partial t}X(t_0;\omega) \quad (\omega \in \Omega).$$

It suffices to prove that $E(X_0)$ is finite, and that

$$\frac{F(t_n)-F(t_0)}{t_n-t_0} \to E(X_0) \quad \text{as} \quad n \to \infty \tag{15}$$

whenever

$$t_1, t_2, \ldots \in I-\{t_0\} \quad \text{and} \quad t_n \to t_0 \text{ as } n \to \infty. \tag{16}$$

Suppose that (16) holds, and let

$$X_n = \frac{X^{(t_n)}-X^{(t_0)}}{t_n-t_0} \quad (n = 1, 2, \ldots),$$

and so $X_1, X_2, \ldots$ are random variables (by Theorems 3.3 (i) and 3.4 (i)). For $n = 1, 2, \ldots$ and $\omega \in \Omega$

$$X_n(\omega) = \frac{\partial}{\partial t} X\{t_0+\theta(t_n-t_0); \omega\},$$

where $0 < \theta = \theta(n, \omega) < 1$ (by the mean value theorem), and so

$$|X_n| \leqslant Y \text{ on } \Omega \qquad (n = 1, 2, \ldots).$$

Now, by (iii), $X_n \to X_0$ on Ω as $n \to \infty$ (and so, by Theorem 3.6 (iii), X_0 is a random variable). The dominated convergence theorem now shows that $E(X_0)$ is finite, and that $E(X_n) \to E(X_0)$ as $n \to \infty$. Thus (15) holds, and the proof is complete.

Note. Theorem 8 still holds if I is not open, i.e. if I is an interval of one of the forms $[a, b]$, $(a, b]$, $[a, b)$, $[a, \infty)$ or $(-\infty, b]$, provided that the derivative at any end point of I which belongs to I is interpreted in the appropriate one-sided sense.

Theorem 9 *Suppose that* (A) *and* (B) *above hold, and suppose also that*

(i) *I is a closed finite interval* $[a, b]$;
(ii) *there exists a random variable Y such that $E(Y)$ is finite and*

$$|X^{(t)}| \leqslant Y \text{ on } \Omega \quad (t \in I);$$

(iii) *for each fixed $\omega \in \Omega$, $X(t; \omega)$ is continuous on I.*

Then $E\left\{\int_a^b X^{(t)}\, dt\right\}$ *is finite and*

$$\int_a^b E\{X^{(t)}\}\, dt = E\left\{\int_a^b X^{(t)}\, dt\right\}. \tag{17}$$

Notes. (a) If $F(t) = E\{X^{(t)}\}$ $(a \leqslant t \leqslant b)$, then F is continuous on $[a, b]$ (by Theorem 7), and so is integrable (in the Riemann sense) over $[a, b]$.

(b) It will appear in the course of the proof that $\int_a^b X^{(t)}\, dt$ is a random variable.

Proof. Let

$$X_n(\omega) = \sum_{r=1}^{n} X\left(a+r\frac{b-a}{n}; \omega\right)\frac{b-a}{n} \qquad (n = 1, 2, \ldots; \omega \in \Omega).$$

Then $X_n(\omega) \to \int_a^b X(t; \omega)\, dt = X_0(\omega)$, say, as $n \to \infty$ $(\omega \in \Omega)$, and so $X_0 =$

$\int_a^b X(t)\,dt$ is a random variable. Now

$$|X_n| \leqslant (b-a)Y \text{ on } \Omega \qquad (n = 1, 2, \ldots).$$

Therefore, by the dominated convergence theorem, $E(X_0)$ is finite and $E(X_n) \to E(X_0)$ as $n \to \infty$. But

$$E(X_n) = \sum_{r=1}^{n} F\left(a+r\frac{b-a}{n}\right)\frac{b-a}{n}$$

$$\to \int_a^b F(t)\,dt \quad \text{as} \quad n \to \infty,$$

because F is continuous on, and therefore integrable over, $[a, b]$. Therefore $\int_a^b F(t)\,dt = E(X_0)$, which is (17).

(IV) Later on it will be convenient, particularly when considering the theory of characteristic functions, to consider the expectations of "complex-valued" random variables.

By a complex-valued random variable is meant a function $X : \Omega \to C$ (where C is the set of all finite complex numbers) of the form

$$X = X' + iX'',$$

where X' and X'' are real-valued random variables. For such a random variable X we define $E(X)$ only when $E(X')$ and $E(X'')$ are both finite.

Definition 4. $E(X)$ is said to be *finite* if $E(X')$ and $E(X'')$ are both finite and then

$$E(X) = E(X') + iE(X'').$$

($E(X')$ and $E(X'')$ are to be interpreted in the sense of Definition 3.)

If X is a real-valued random variable, it is easy to see that Definitions 3 and 4 give the same value of $E(X)$.

The theorems which still have a meaning if complex-valued random variables are admitted, namely Theorems 1, 2, 4, 6, 7, 8 and 9, and their corollaries, continue to hold. (In Theorem 1, c can be a complex number; Y is real-valued, and, in fact, non-negative, in Theorems 4 (iii), 6, 7, 8 and 9.)

With the exception of Theorem 4 (i) and (ii), the proofs are straightforward, and follow from the previous results by considering the real and imaginary parts separately. The details are left to the reader.

When we refer to the "complex-valued" form of a theorem in what follows, it will be denoted by the affix C, e.g. Theorem 1C.

Proof of Theorem 4 (i) *and* (ii). (i) This follows from the inequalities

$$|X'|,\ |X''| \leqslant |X| \leqslant |X'|+|X''| \text{ on } \Omega,$$

on noting that $|X| = \sqrt{\{(X')^2+(X'')^2\}}$ is a random variable (by Theorems 3.3 and 3.4).

(ii) Let $c = E(X)$. It may be assumed that $c \neq 0$, and now

$$|c| = E\left(\frac{|c|}{c}X\right) \quad \text{(by Theorem 1C),}$$

i.e. $$|c| = E\left\{Re\left(\frac{|c|}{c}X\right)\right\}+iE\left\{Im\left(\frac{|c|}{c}X\right)\right\} \quad \text{(by Definition 4).}$$

The left-hand side is real, and so the second term on the right must be zero. Also

$$Re\left(\frac{|c|}{c}X\right) \leqslant \left|\frac{|c|}{c}X\right| = |X| \text{ on } \Omega.$$

Therefore

$$E\left\{Re\left(\frac{|c|}{c}X\right)\right\} \leqslant E(|X|) \quad \text{(by Theorem 3),}$$

and the result follows.

To prepare for our next theorem we remind the reader that the distribution of a random variable X is the function $P_X\colon \mathscr{B} \to [0, 1]$ defined by

$$P_X(B) = P\{X^{-1}(B)\} \quad (B \in \mathscr{B}), \tag{18}$$

and that $(R, \mathscr{B}, P_X)$ is a probability space (see Exercise 4.1 (i)). In that theorem it will be shown that the existence of $E(X)$ and its value, if it exists, depend only on the distribution P_X of X. An equivalent assertion is that if X and Y are two random variables for which $P_X = P_Y$ on $\mathscr{B}$, then $E(X)$ exists if and only if $E(Y)$ exists, and in that case $E(X) = E(Y)$. (The probability spaces on which X and Y are defined may differ.)

Let I be the identity function on R, i.e. the function $I: R \to R$ defined by

$$I(x) = x \quad (x \in R).$$

Clearly I is Borel measurable, and so is a random variable defined on the probability space $(R, \mathscr{B}, P_X)$.

In part (i) of the following theorem we shall be concerned both with $E(X)$, i.e. the expectation of the random variable X defined on the probability space $(\Omega, \mathscr{F}, P)$, and with $E(I)$, i.e. the expectation of the random variable I defined on the probability space $(R, \mathscr{B}, P_X)$. The integral notation mentioned after Definition 3, which makes explicit reference to the sample space and probability measure involved, helps us to keep in mind the different probability spaces which occur. With that notation

$$E(X) = \int_{\Omega} X(\omega)\, dP$$

and

$$E(I) = \int_{R} I(x)\, dP_X = \int_{R} x\, dP_X.$$

Corresponding remarks apply in part (ii) of the theorem.

Theorem 10 *With the above notation*

(i) *$E(X)$ exists if and only if $E(I)$ exists, and then*

$$E(X) = E(I), \quad i.e. \quad \int_{\Omega} X(\omega)\, dP = \int_{R} x\, dP_X;$$

(ii) *if also $g: R \to R$ is Borel measurable, $E\{g(X)\}$ exists if and only if $E\{g(I)\}$ exists, and then*

$$E\{g(X)\} = E\{g(I)\}, \quad i.e. \quad \int_{\Omega} g\{X(\omega)\}\, dP = \int_{R} g(x)\, dP_X.$$

Notes. (a) By Theorem 3.5, $g(X)$ is a random variable defined on $(\Omega, \mathscr{F}, P)$ and $g(I)$ is a random variable defined on $(R, \mathscr{B}, P_X)$.

(b) It follows from (ii) that, for given g, the existence of $E\{g(X)\}$ and its value, if it exists, are determined solely by the distribution P_X of X, and therefore by the distribution function of X (see Theorem 4.2, corollary 2).

Proof. It suffices to prove (ii), since (i) is a special case $(g(x) = x)$ of (ii).

The proof proceeds in three stages, corresponding to the three stages in the definition of an expected value (in this theorem we shall assume that both X and g are real-valued).

(I) Suppose firstly that g is simple and non-negative on R. Let the distinct values taken by g be $a_1, a_2, \ldots, a_m$. For $i = 1, 2, \ldots, m$ let

$$A_i = \{x : g(x) = a_i\} \qquad (\subseteq R)$$

and

$$B_i = \{\omega : g\{X(\omega)\} = a_i\} \quad (\subseteq \Omega),$$

and so

$$A_i \in \mathscr{B}$$

and

$$B_i = \{\omega : X(\omega) \in A_i\} = X^{-1}(A_i) \in \mathscr{F}.$$

Also

$$g(x) = \sum_{i=1}^{m} a_i I_{A_i}(x) \quad (x \in R)$$

and

$$g\{X(\omega)\} = \sum_{i=1}^{m} a_i I_{B_i}(\omega) \quad (\omega \in \Omega).$$

Therefore

$$\begin{aligned} E\{g(X)\} &= \sum_{i=1}^{m} a_i P(B_i) && \text{(by Definition 1)} \\ &= \sum_{i=1}^{m} a_i P\{X^{-1}(A_i)\} \\ &= \sum_{i=1}^{m} a_i P_X(A_i) && \text{(by (18))} \\ &= E\{g(I)\} && \text{(by Definition 1).} \end{aligned}$$

(II) Suppose secondly that g is non-negative on R. Then there exist simple Borel measurable functions $g_1, g_2, \ldots$ such that

$$0 \leqslant g_1 \leqslant g_2 \leqslant \ldots \text{ on } R$$

and

$$g_n \to g \text{ on } R \quad \text{as} \quad n \to \infty.$$

Therefore, by Definition 2,

$$E\{g_n(I)\} \to E\{g(I)\} \quad \text{as} \quad n \to \infty.$$

Now $g_1(X), g_2(X)$ are simple $\mathscr{F}$-measurable functions satisfying

$$0 \leqslant g_1(X) \leqslant g_2(X) \leqslant \ldots \text{ on } \Omega$$

and

$$g_n(X) \to g(X) \text{ on } \Omega \quad \text{as} \quad n \to \infty.$$

Therefore, again by Definition 2,

$$E\{g_n(X)\} \to E\{g(X)\} \quad \text{as} \quad n \to \infty.$$

Since, by (I),

$$E\{g_n(X)\} = E\{g_n(I)\} \qquad (n = 1, 2, \ldots),$$

the result follows on letting n tend to infinity.

(III) Suppose finally that g can take values of either sign on R. Then g^+ and g^- are non-negative Borel measurable functions, and so, by (II),

$$E\{g^+(X)\} = E\{g^+(I)\}$$

and

$$E\{g^-(X)\} = E\{g^-(I)\}.$$

Since $g^+(X) = \{g(X)\}^+$ and $g^-(X) = \{g(X)\}^-$ on Ω, the result follows by Definition 3.

Evaluation of $E(X)$ in two special cases

Although $E(X)$ has now been defined, the definition is somewhat abstract and does not give a convenient method of calculating $E(X)$ (unless X is a simple random variable). In what follows, we show how to determine whether $E(X)$ exists, and its value if it does exist, in two particularly important cases.

Case 1. X is a *discrete* random variable, i.e. X can take only a countable number of distinct values (which is certainly the case if the sample space Ω is countable). If X takes only a finite number of distinct values it is simple, and there is no problem in evaluating $E(X)$. We shall therefore assume that X takes an enumerable infinity of distinct values.

Case 2. X is an *absolutely continuous* random variable, i.e. X has a *probability density function.* This is a Borel measurable function $f: R \to [0, \infty)$ with the property that

$$P_X(A) = P\{X^{-1}(A)\} = \int_A f(x)\,dx \tag{19}$$

for every Borel set A, and so, in particular, $\int_R f(x)\,dx = 1.$

The integral is to be interpreted in the Lebesgue sense, any Borel measurable function being also Lebesgue measurable (see Appendix A). However, although Lebesgue integrals are used to develop the general theory, the probability density functions of those random variables which are of practical importance have only a finite number of points of discontinuity, and the integration is usually to be effected over some interval (finite or infinite). Thus the integrals which occur in practice can be interpreted in the Riemann or Cauchy–Riemann sense, and evaluated by elementary methods.

If X is an absolutely continuous random variable then

$$P(X = c) = 0 \quad (c \in R),$$

and so the distribution function of X is everywhere continuous (see Exercise 4.2). Unfortunately, the converse statement is false. It is possible to construct random variables whose distribution functions are everywhere continuous but which do not have probability density functions. Such random variables (continuous, but not absolutely continuous) have only a pathological significance; they are of no practical importance.

It should, however, be mentioned, that there do exist random variables of practical importance whose distribution functions can best be described as a mixture of the discrete and the absolutely continuous. For example, let X be the time spent waiting for service in a post office. Then there is a positive probability p that $X = 0$, whereas

$$P(a < X < b) = \int_a^b f(x)\,dx \quad (0 \leqslant a < b \leqslant \infty)$$

with some suitable $f(x)$; in particular $\int_0^\infty f(x)\,dx = 1-p$. The expected value of such a random variable can be evaluated by a suitable combination of the methods we are about to describe.

Case 1. X is a discrete random variable. If X is a discrete random variable, there exist disjoint sets $A_1, A_2, \ldots$ of $\mathscr{F}$ such that

$$X \text{ is a constant } x_n, \text{ say, on } A_n \qquad (n = 1, 2, \ldots)$$

and

$$\bigcup_{n=1}^{\infty} A_n = \Omega.$$

(We do not necessarily assume that $x_1, x_2, \ldots$ are distinct.) Let

$$p_n = P(A_n) \qquad (n = 1, 2, \ldots).$$

Then $E(X)$ is finite if and only if $\sum_{n=1}^{\infty} p_n x_n$ is absolutely convergent, and then

$$E(X) = \sum_{n=1}^{\infty} p_n x_n .$$

Proof. We suppose firstly that $X \geqslant 0$ on Ω, and let

$$X_k = XI_{A_1} + XI_{A_2} + \ldots + XI_{A_k} = \sum_{n=1}^{k} x_n I_{A_n} \qquad (k = 1, 2, \ldots).$$

Then $X_1, X_2, \ldots$ are simple random variables satisfying

$$0 \leqslant X_1 \leqslant X_2 \leqslant \ldots \text{ on } \Omega$$

and

$$X_k \to X \text{ on } \Omega \quad \text{as} \quad k \to \infty.$$

Now

$$E(X_k) = \sum_{n=1}^{k} p_n x_n \qquad (k = 1, 2, \ldots),$$

and so, by Definition 2,

$$E(X) = \sum_{n=1}^{\infty} p_n x_n , \tag{20}$$

where the right-hand side is to be interpreted as ∞ if the series is divergent.

In particular, $E(X)$ is finite if and only if the series $\sum_{n=1}^{\infty} p_n x_n$ is convergent (and therefore absolutely convergent, because all its terms are non-negative).

Suppose secondly that X can take values of either sign on Ω. Then $E(X)$ is finite if and only if $E(|X|)$ is finite (by Theorem 4 (i)), i.e. if and only if $\sum_{n=1}^{\infty} p_n |x_n|$ is convergent (by what has just been proved), i.e. if and only if $\sum_{n=1}^{\infty} p_n x_n$ is absolutely convergent. If $\sum_{n=1}^{\infty} p_n x_n$ is absolutely convergent

$$\begin{aligned} E(X) &= E(X^+) - E(X^-) \qquad \text{(by Definition 3)} \\ &= \sum_{n=1}^{\infty} p_n x_n^+ - \sum_{n=1}^{\infty} p_n x_n^- \qquad \text{(see (20))} \\ &= \sum_{n=1}^{\infty} p_n x_n . \end{aligned}$$

A slight modification of the argument shows that $E(X)$ exists (as opposed to being finite) if and only if at least one of the series

$$\sum_{n=1}^{\infty} p_n x_n^+ \quad \text{and} \quad \sum_{n=1}^{\infty} p_n x_n^-$$

is convergent, and then

$$E(X) = \sum_{n=1}^{\infty} p_n x_n$$

as before.

It follows immediately that for any function $g : R \to R$ (not necessarily Borel measurable) $E\{g(X)\}$ exists if and only if at least one of the series

$$\sum_{n=1}^{\infty} p_n g^+(x_n) \quad \text{and} \quad \sum_{n=1}^{\infty} p_n g^-(x_n)$$

is convergent, and then

$$E\{g(X)\} = \sum_{n=1}^{\infty} p_n g(x_n).$$

In particular, $E\{g(X)\}$ is finite if and only if $\sum_{n=1}^{\infty} p_n g(x_n)$ is absolutely convergent.

Case 2. X is an absolutely continuous random variable. Suppose that X has a probability density function f with the properties specified above. Then

(i) $E(X)$ is finite if and only if $\int_R xf(x)\,dx$ is finite, and then

$$E(X) = \int_R xf(x)\,dx;$$

(ii) if also $g : R \to R$ is Borel measurable, $E\{g(X)\}$ is finite if and only if $\int_R g(x)f(x)\,dx$ is finite, and then

$$E\{g(X)\} = \int_R g(x)f(x)\,dx.$$

Note. The integrals are to be interpreted in the Lebesgue sense, and consequently $\int_R g(x)f(x)\,dx$ is finite if and only if $\int_R |g(x)|\,f(x)\,dx$ is finite. In problems of practical importance, $f(x)$ and $g(x)$ are "well-behaved" and $\int_R g(x)f(x)\,dx$ can be evaluated as the Cauchy–Riemann integral $\int_{-\infty}^{\infty} g(x)f(x)\,dx$. However, the reader must bear in mind that a Cauchy–Riemann integral $\int_{-\infty}^{\infty} h(x)\,dx$ may be finite when the corresponding Lebesgue integral $\int_R h(x)\,dx$ does not exist; this occurs when the Cauchy–Riemann integral is not absolutely convergent, e.g. when $h(x) = (\sin x)/x$. Thus with a "well-behaved" function $h(x)$, e.g. one having only a finite number of points of discontinuity, the Lebesgue integral $\int_R h(x)\,dx$ is finite if and only if the Cauchy–Riemann integral $\int_{-\infty}^{\infty} h(x)\,dx$ is absolutely convergent, and the two integrals then have the same value.

Proof. It suffices to prove (ii).

(I) Suppose firstly that g is simple and non-negative on R. Let the distinct values of g be $a_1, a_2, \ldots, a_m$, and define A_i and B_i for $i = 1, 2, \ldots, m$ as in the proof of Theorem 10. Then

$$\begin{aligned} E\{g(X)\} &= \sum_{i=1}^{m} a_i P(B_i) \\ &= \sum_{i=1}^{m} a_i P\{X^{-1}(A_i)\} \end{aligned}$$

$$= \sum_{i=1}^{m} a_i \int_{A_i} f(x)\,dx \qquad \text{(by (19))}$$

$$= \sum_{i=1}^{m} \int_{A_i} g(x)f(x)\,dx \qquad \text{(because } g(x) = a_i \text{ on } A_i)$$

$$= \int_R g(x)f(x)\,dx,$$

because $A_1, A_2, \ldots, A_m$ are disjoint sets of $\mathscr{B}$ whose union is R (see Appendix A).

(II) Suppose secondly that g is non-negative on R. Let $g_1, g_2, \ldots$ be defined as in the proof of Theorem 10. Then, by (I),

$$E\{g_n(X)\} = \int_R g_n(x)f(x)\,dx \qquad (n = 1, 2, \ldots).$$

As $n \to \infty$

$$E\{g_n(X)\} \to E\{g(X)\} \qquad \text{(by Definition 2)}$$

and

$$\int_R g_n(x)f(x)\,dx \to \int_R g(x)f(x)\,dx$$

(by the monotone convergence theorem for the Lebesgue integral). The result follows.

(III) Suppose finally that g can take values of either sign on R. Then, by (II)'

$$E\{g^+(X)\} = \int_R g^+(x)f(x)\,dx$$

and

$$E\{g^-(X)\} = \int_R g^-(x)f(x)\,dx.$$

The result follows since $\int_R g(x)f(x)\,dx$ is finite if and only if $\int_R |g(x)|f(x)\,dx$ is finite, and this holds if and only if $\int_R g^+(x)f(x)\,dx$ and $\int_R g^-(x)f(x)\,dx$ are both finite.

It is left to the reader to modify the above argument so as to determine whether $E\{g(X)\}$ exists, and its value if it does exist.

The expression of $E(X)$ as a Riemann–Stieltjes integral

The methods just described for determining whether $E(X)$ exists, and evaluating it if it does exist, are limited to random variables which are either discrete or absolutely continuous (i.e. possess a probability density function). Though these methods are both convenient and readily applicable, it does seem inappropriate to use the relatively formidable machinery of Lebesgue integration to evaluate $E(X)$ in those practically important cases when X is an absolutely continuous random variable whose probability density function has only a finite number of points of discontinuity.

We shall end this chapter by obtaining a further expression for $E(X)$, namely one in terms of Riemann–Stieltjes integrals, which has certain advantages. Firstly, the same expression holds for random variables of all kinds, be they discrete or absolutely continuous (or neither). This, of course, is true of the expression $E(X)$ itself, but certain arguments are more naturally handled in terms of Riemann–Stieltjes integrals. Secondly, when X has a probability density function with only a finite number of points of discontinuity, the value of $E(X)$ is obtained as a Cauchy–Riemann integral without the lengthy detour via the corresponding Lebesgue integral.

The same remarks apply to $E\{g(X)\}$, where $g : R \to R$ is continuous on R.

Thus the Riemann–Stieltjes integral will be particularly appropriate when X is an absolutely continuous random variable having a sufficiently well-behaved probability density function, and when we are interested in $E(X)$ itself or, more generally, in $E\{g(X)\}$ for some continuous function g (as opposed to one which is only Borel measurable). For it is precisely such absolutely continuous random variables and expected values which occur in practice, and then the Riemann–Stieltjes integral leads immediately and naturally to the corresponding Cauchy–Riemann integral.

We give a brief account of the Riemann–Stieltjes integral in Appendix C for the reader who is not familiar with that topic.

In what immediately follows we shall suppose that X is a random variable with distribution function F, and that $g : R \to R$ is continuous on R.

Lemma 4. *Suppose that* $Y = g(X)$ *if* $a < X \leqslant b$ *and* $Y = 0$ *otherwise (and so* Y *is a bounded random variable). Then*

$$E(Y) = \int_a^b g\, dF.$$

Note. $E(Y)$ is finite (by Theorem 4, corollary), and $\int_a^b g\, dF$ exists (by Theorem C.5).

Proof. Let $x_r = a + r\dfrac{b-a}{n}$ $(r = 0, 1, \ldots, n)$. Let also $M_r = M_r(n)$ and $m_r = m_r(n)$ denote respectively the upper and lower bounds of g on $[x_{r-1}, x_r]$ $(r = 1, 2, \ldots, n)$. Then

$$m_r \leqslant g(X) = Y \leqslant M_r \quad \text{if} \quad x_{r-1} < X \leqslant x_r \qquad (r = 1, 2, \ldots, n).$$

Therefore, since

$$P(x_{r-1} < X \leqslant x_r) = F(x_r) - F(x_{r-1}) \qquad (r = 1, 2, \ldots, n),$$

$$\sum_{r=1}^{n} m_r\{F(x_r) - F(x_{r-1})\} \leqslant E(Y) \leqslant \sum_{r=1}^{n} M_r\{F(x_r) - F(x_{r-1})\},$$

or, say,

$$s_n \leqslant E(Y) \leqslant S_n.$$

Now also

$$s_n \leqslant \sum_{r=1}^{n} g(x_r)\{F(x_r) - F(x_{r-1})\} \leqslant S_n,$$

or, say

$$s_n \leqslant \Sigma_n \leqslant S_n,$$

and so

$$|\Sigma_n - E(Y)| \leqslant S_n - s_n = \sum_{r=1}^{n} (M_r - m_r)\{F(x_r) - F(x_{r-1})\}.$$

Since

$$\Sigma_n \to \int_a^b g\, dF \quad \text{as} \quad n \to \infty$$

and

$$\max\{M_r - m_r : r = 1, 2, \ldots, n\} \to 0 \quad \text{as} \quad n \to \infty$$

(because g is continuous on $[a, b]$), the result follows.

Theorem 11 *Suppose that X is a random variable with distribution function F, and that $g : R \to R$ is continuous on R. Then $E\{g(X)\}$ is finite if and only if $\int_{-\infty}^{\infty} g\, dF$ is absolutely convergent, and in that case*

$$E\{g(X)\} = \int_{-\infty}^{\infty} g\, dF. \tag{21}$$

Proof. For $n = 1, 2, \ldots$ let

$$Y_n = \begin{cases} g(X) & \text{if} \quad -n < X \leqslant n \\ 0 & \text{otherwise.} \end{cases}$$

Then, by Lemma 4,

$$E(|Y_n|) = \int_{-n}^{n} |g|\, dF.$$

Now if $E\{g(X)\}$ is finite, the numbers $E(|Y_n|)$ $(n = 1, 2, \ldots)$ are bounded above (by $E\{|g(X)|\}$), and so the numbers $\int_{-n}^{n} |g|\, dF$ $(n = 1, 2, \ldots)$ are bounded above, from which it follows that $\int_{-\infty}^{\infty} g\, dF$ is absolutely convergent.

Conversely, if $\int_{-\infty}^{\infty} g\, dF$ is absolutely convergent,

$$E(|Y_n|) = \int_{-n}^{n} |g|\, dF \to \int_{-\infty}^{\infty} |g|\, dF \quad \text{as} \quad n \to \infty.$$

Since also

$$E(|Y_n|) \to E\{|g(X)|\} \quad \text{as} \quad n \to \infty$$

(by the monotone convergence theorem), $E\{|g(X)|\}$ is finite, and so $E\{g(X)\}$ is finite.

Finally, to show that (21) holds, note that

$$E(Y_n) = \int_{-n}^{n} g\, dF \quad \text{(by the lemma)}$$

$$\to \int_{-\infty}^{\infty} g\, dF \quad \text{as} \quad n \to \infty,$$

and

$$E(Y_n) \to E\{g(X)\} \quad \text{as} \quad n \to \infty$$

(by the dominated convergence theorem).

In particular, it follows from the theorem that $E(X)$ is finite if and only if $\int_{-\infty}^{\infty} x\, dF(x)$ is absolutely convergent, and then

$$E(X) = \int_{-\infty}^{\infty} x\, dF(x).$$

This shows once again that whether $E(X)$ is finite, and its value if it is finite, depend solely on the distribution function of X.

Now let us suppose that X has a probability density function with only a finite number of points of discontinuity, i.e. that there exists a function $f: R \to [0, \infty)$ with only a finite number of points of discontinuity for which

$$F(x) = \int_{-\infty}^{x} f(t)\, dt \quad (x \in R)$$

(see Exercise 25). Since $F'(x) = f(x)$ for all real x with at most a finite number of exceptions, it follows from Theorems C.4 and C.7, corollary, that $E\{g(X)\}$

is finite if and only if the Cauchy–Riemann integral $\int_{-\infty}^{\infty} g(x)f(x)\,dx$ is absolutely convergent, and then

$$E\{g(X)\} = \int_{-\infty}^{\infty} g(x)f(x)\,dx.$$

This formula and that given earlier for the expected value of a discrete random variable will enable us to calculate the expected values of most random variables which are of practical importance.

As an example of the use of Riemann–Stieltjes integrals, we show finally that $E(X)$ is finite if and only if the Riemann integrals

$$\int_{0}^{\infty} \{1-F(x)\}\,dx \quad \text{and} \quad \int_{-\infty}^{0} F(x)\,dx \tag{22}$$

are convergent, and then

$$E(X) = \int_{0}^{\infty} \{1-F(x)\}\,dx - \int_{-\infty}^{0} F(x)\,dx. \tag{23}$$

Suppose firstly that $E(X)$ is finite. Then $\int_{-\infty}^{\infty} x\,dF(x)$ is absolutely convergent, and so

$$\int_{0}^{\infty} x\,dF(x), \quad \int_{-\infty}^{0} x\,dF(x) \tag{24}$$

are both convergent. Now, for $b > 0$,

$$\begin{aligned}\int_{0}^{b} x\,dF(x) &= -\int_{0}^{b} x\,d\{1-F(x)\}\\ &= -b\{1-F(b)\}+\int_{0}^{b} \{1-F(x)\}\,dx \end{aligned}\tag{25}$$

(by theorem C.3), and

$$\int_{-b}^{0} x\,dF(x) = bF(-b)-\int_{-b}^{0} F(x)\,dx \tag{26}$$

(again by theorem C.3). Since

$$\begin{aligned} bF(-b)+b\{1-F(b)\} &\leqslant bP(|X| \geqslant b)\\ &\to 0 \quad \text{as} \quad b \to \infty \quad \text{(by Exercise 18 (ii))}, \end{aligned}$$

it follows that the integrals (22) are convergent.

Conversely, suppose that the integrals (22) are convergent. It follows from (25) and (26) that

$$\int_0^b x\,dF(x) \leqslant \int_0^b \{1-F(x)\}\,dx$$

and

$$\int_{-b}^0 x\,dF(x) \geqslant -\int_{-b}^0 F(x)\,dx,$$

and hence that the integrals (24) are both convergent, i.e. that $\int_{-\infty}^{\infty} x\,dF(x)$ is absolutely convergent, and so $E(X)$ is finite.

Finally, (23) follows on letting b tend to infinity in (25) and (26) and noting that

$$bF(-b), \quad b\{1-F(b)\} \to 0 \quad \text{as} \quad b \to \infty.$$

Two definitions

We give here definitions of two terms which will be used later in this book, in particular in the exercises at the end of this chapter.

(1) The phrase "*almost surely*" (abbreviated to a.s.) means "except possibly on a set of probability zero". Thus to say that the random variable X vanishes a.s. means that there exists a set A of $\mathscr{F}$ such that $P(A) = 0$ and $X = 0$ on A^c. Similarly, if X and Y are two random variables, to say that $X = Y$ a.s. means that there is a set A of $\mathscr{F}$ such that $P(A) = 0$ and $X = Y$ on A^c. It is almost immediate that if $X = Y$ a.s. then X and Y have the same distribution function. See also note (1) on page 34.

(2) If X is a random variable for which $E(X^2)$ is finite then it can be shown that $E(X)$ is also finite (see Exercise 4), and the *variance* of X (var X) is defined to be

$$E[\{X-E(X)\}^2].$$

The requirement that $E(X^2)$ be finite suffices to ensure that var X is finite and equal to

$$E(X^2)-\{E(X)\}^2,$$

from which it follows that $\{E(X)\}^2 \leqslant E(X^2)$ (see also Exercise 6).

Exercises

1. Suppose that the random variable X vanishes a.s. Prove that $E(X) = 0$

   (i) by use of Definitions 1, 2 and 3;
   (ii) by use of Theorem 11;
   (iii) by use of equation (23).

2. (i) Suppose that $X \geqslant 0$ on Ω, and that $E(X) = 0$. Prove that $X = 0$ a.s.

Hint. Let $A_n = \left\{X > \frac{1}{n}\right\}$ $(n = 1, 2, \ldots)$. Then, for $n = 1, 2, \ldots,$

$$E(X) \geqslant E(XI_{A_n}) \geqslant \frac{1}{n} P(A_n), \quad \text{and so} \quad P(A_n) = 0.$$

Note that $\{X > 0\} = \bigcup_{n=1}^{\infty} A_n$.

(ii) Let Y be a non-negative random variable satisfying $P(Y > 0) > 0$. Prove that $E(Y) > 0$.

3. Suppose that $E(X)$ is finite, and that $A \in \mathscr{F}$. By using Theorem 4, or otherwise, prove that $E(XI_A)$ is finite.

4. Suppose that X is a random variable, and that $E(X^n)$ is finite, where n is a positive integer ($E(X^n)$ is the *nth moment* of X). Prove that $m_r = E(|X|^r)$ is finite for every real number r in $(0, n]$, and deduce that $E(X^k)$ is finite ($k = 1, 2, \ldots, n$).

Hint. $|X|^r \leqslant 1 + |X|^n$ on Ω; refer to Exercise 3.5.

Prove also that

$$m_r^{1/r} \leqslant m_s^{1/s} \quad (0 < r < s \leqslant n).$$

Hint. Let $c = \frac{s}{r} (> 1)$. The curve $y = x^c$ $(x \geqslant 0)$ is concave upwards, and so

$$|X|^s \geqslant c m_r^{c-1}(|X|^r - m_r) + m_r^c \text{ on } \Omega.$$

5. Suppose that $E(X)$ is finite, and let Y be a random variable such that $Y = X$ a.s.

(a) Prove that X and Y have the same distribution, and the same distribution function.

(b) Prove that $E(Y) = E(X)$

(i) by use of Exercise 1 and Theorem 2;
(ii) by use of Theorem 10;
(iii) by use of Theorem 11.

6. Suppose that X and Y are random variables with finite second moments.

(a) Prove that $E(XY)$ is finite.

Hint. $|XY| \leqslant \frac{1}{2}(X^2 + Y^2)$.

(b) Prove that

$$\{E(XY)\}^2 \leqslant E(X^2)E(Y^2)$$

(the *Cauchy–Schwarz inequality*), and that equality holds if and only if there exist real numbers s and t which are not both zero such that

$$sX+tY = 0 \quad \text{a.s.}$$

Hint. If $a = E(X^2)$, $h = E(XY)$, $b = E(Y^2)$ then

$$as^2+2hst+bt^2 = E\{(sX+tY)^2\} \quad (s, t \in R)$$

c) Prove that

$$|\sqrt{\text{var } X}-\sqrt{\text{var } Y}| \leqslant \sqrt{\text{var } (X \pm Y)} \leqslant \sqrt{\text{var } X}+\sqrt{\text{var } Y}.$$

7. (i) Prove that var $X = 0$ if and only if X is a.s. equal to a constant $(= E(X))$.

(ii) Let X and Y be two random variables with finite second moments, and suppose that

$$\text{var } X > 0, \quad \text{var } Y > 0.$$

The *coefficient of correlation* (ϱ) of X and Y is then defined to be

$$\frac{E[\{X-E(X)\}\{Y-E(Y)\}]}{\sqrt{\{(\text{var } X)(\text{var } Y)\}}}.$$

Prove that

(a) $|\varrho| \leqslant 1$;

(b) $|\varrho| = 1$ if and only if there exist real numbers a and b with $a \neq 0$ such that $Y = aX+b$ a.s., and then $\varrho = a/|a|$.

Hint. Use Exercise 6.

8. Prove *Chebyshev's inequality*, namely that if X is a random variable with finite second moment then

$$P(|X-E(X)| \geqslant t) \leqslant (\text{var } X)/t^2 \quad (t > 0).$$

Hint. Let $Y = |X-E(X)|$ and $A = \{Y \geqslant t\}$. Then

$$E(Y^2) \geqslant E(Y^2 I_A) \geqslant t^2 P(A).$$

9. Suppose that X is a random variable with finite second moment, and that var $X > 0$. Prove that

$$P(|X-E(X)| > t) < (\text{var } X)/t^2 \quad (t > 0).$$

(a strict form of Chebyshev's inequality).

Hint. Let $Y = |X-E(X)|$ and $A = \{Y > t\}$. Then, if $P(A) > 0$,

$$E(Y^2) \geqslant E(Y^2 I_A) > t^2 P(A),$$

the second inequality following from Exercise 2(ii).

Note. There are really four forms of Chebyshev's inequality, that given in Exercise 8 being the basic one. As they are not infrequently confused, we list them below. Assume throughout that X is a random variable with finite second moment, and let $Y = |X-E(X)|$, $\sigma = \sqrt{\operatorname{var} X}$. Then for $t > 0$

(i) (a) $P(Y \geqslant t) \leqslant \sigma^2/t^2$ (Exercise 8);
(b) $P(Y > t\sigma) < 1/t^2$;

(ii) if also $\sigma > 0$

(a) $P(Y > t) < \sigma^2/t^2$ (Exercise 9);
(b) $P(Y \geqslant t\sigma) \leqslant 1/t^2$.

It is left to the reader to obtain the inequalities (i) (b) and (ii) (b), and satisfy himself of the need for the condition $\sigma > 0$ in (ii).

10. (i) Prove that if f is non-decreasing on $[0, \infty)$ and $f(0) \geqslant 0$ then

$$E\{f(|X|)\} \geqslant f(t)\, P(|X| \geqslant t) \quad (t > 0)$$

for any random variable X.

Deduce Chebyshev's inequality.

(ii) If also $f \leqslant M$ on $[0, \infty)$, or even $f(|X|) \leqslant M$ on Ω (which is a weaker condition), prove that

$$E\{f(|X|)\} \leqslant f(t) + \{M - f(t)\}\, P(|X| \geqslant t) \quad (t > 0).$$

Hints. Let $A = \{|X| \geqslant t\}$. Then (i) $f(|X|) \geqslant f(t) I_A$;
(ii) $f(|X|) \leqslant f(t)(1-I_A) + M I_A$.

11. Suppose that X is a random variable. Prove that

$$P(X \geqslant t) \leqslant e^{-ct} E(e^{cX})$$

for any positive number c and any real number t.

12. Suppose that $E(X) = 0$, $\operatorname{var} X = 1$, and $|X| \leqslant M$ on Ω. Prove that

$$P(|X| \geqslant t) \geqslant (1-t^2)/(M^2-t^2) \quad (0 < t < 1).$$

13. Suppose that c is a real number, and that $X_1, X_2, \ldots$ are random variables satisfying

$$E(X_n^2) < \infty \qquad (n = 1, 2, \ldots)$$

and

$$E\{(X_n-c)^2\} \to 0 \quad \text{as} \quad n \to \infty.$$

Prove that

(i) $E(X_n) \to c$ as $n \to \infty$;
(ii) for any $\varepsilon > 0$,

$$P(|X_n-c| \geqslant \varepsilon) \to 0 \quad \text{as} \quad n \to \infty.$$

14. Suppose that X is a random variable with finite second moment, and let m be any median of X (see Exercise 4.9). Prove that

$$|m-E(X)| \leqslant \sqrt{(2 \operatorname{var} X)}.$$

Hint. Suppose that $m > E(X)+\sqrt{(2 \operatorname{var} X)}$, and use a suitable form of Chebyshev's inequality.

15. The random variables $X_1, X_2, \ldots$ are non-negative, and the sequence

$$X_1(\omega),\; X_2(\omega),\; \ldots$$

is bounded for each $\omega \in \Omega$. Prove that

$$\varliminf_{n\to\infty} E(X_n) \geqslant E\left(\varliminf_{n\to\infty} X_n\right)$$

(*Fatou's lemma*).

Hint. Apply the monotone convergence theorem to the sequence $Y_1, Y_2, \ldots$, where $Y_n = \inf(X_n, X_{n+1}, \ldots)$ $(n = 1, 2, \ldots)$.

16. The random variables $X_1, X_2, \ldots$ are non-negative, and $\sum_{n=1}^{\infty} X_n$ is convergent on Ω. Prove that

$$E\left(\sum_{n=1}^{\infty} X_n\right) = \sum_{n=1}^{\infty} E(X_n).$$

Hint. Use the monotone convergence theorem.

17. Suppose that $E(X)$ is finite, and let $A_n = \{|X| \leqslant n\}$ $(n = 1, 2, \ldots)$. Prove that $\sum_{n=1}^{\infty} n^{-2}E(X^2 I_{A_n})$ is convergent.

Hint. Let $Y = \sum_{n=1}^{\infty} n^{-2}X^2 I_{A_n}$. Then for $k-1 < |X| \leqslant k$, where k is a positive integer,

$$Y = X^2 \sum_{n=k}^{\infty} n^{-2} \leqslant \frac{2}{k} X^2 \leqslant 2|X|.$$

18. Let X be any random variable, and let

$$A_n = \{n \leqslant |X| < n+1\} \qquad (n = 0, 1, 2, \ldots).$$

Prove that

$$\sum_{n=1}^{\infty} P(|X| \geqslant n) = \sum_{n=1}^{\infty} nP(A_n) \leqslant E(|X|)$$

$$\leqslant \sum_{n=0}^{\infty} (n+1)\, P(A_n)$$

$$= 1 + \sum_{n=1}^{\infty} P(|X| \geqslant n).$$

Hint. $|X| = \sum_{n=0}^{\infty} |X|\, I_{A_n}$; use Exercise 16.

Deduce that

(i) $E(X)$ is finite if and only if $\sum_{n=1}^{\infty} P(|X| \geqslant n)$ is convergent;

(ii) if $E(X)$ is finite then

$$nP(|X| \geqslant n) \to 0 \quad \text{as} \quad n \to \infty,$$

and, more generally,

$$tP(|X| \geqslant t) \to 0 \quad \text{as} \quad t \to \infty$$

(where t tends to infinity through real values).

Hint. Note that $nP(|X| \geqslant n) = \sum_{m=n}^{\infty} nP(A_m) \leqslant \sum_{m=n}^{\infty} mP(A_m)$.

19. Suppose that X is a random variable.

(i) Show that if $E(|X|^r)$ is finite, where $r > 0$, then

$$t^r P(|X| \geqslant t) \to 0 \qquad \text{as} \quad t \to \infty.$$

(ii) Show that if

$$t^r P(|X| \geqslant t) = O(1) \quad \text{as} \quad t \to \infty$$

then $E(|X|^s)$ is finite for $0 < s < r$.

(iii) Construct a random variable X for which $tP(|X| \geqslant t) \to 0$ as $t \to \infty$ and $E(|X|) = \infty$.

Hint. Try $P(X = n) = c/(n^2 \log n)$ $(n = 2, 3, \ldots)$, where c is a suitable constant.

20. (Extension of Lemma 2) Suppose that $X_1, X_2, \ldots$ and Y are non-negative random variables satisfying

$$0 \leqslant X_1 \leqslant X_2 \leqslant \ldots \text{ on } \Omega$$

and

$$0 \leqslant Y \leqslant \lim_{n\to\infty} X_n \text{ on } \Omega.$$

Prove that

$$\lim_{n\to\infty} E(X_n) \geqslant E(Y).$$

Hint. Consider the random variables $Z_n = \min(Y, X_n)$ $(n = 1, 2, \ldots)$, and apply the monotone convergence theorem.

21. Let $A_1, A_2, \ldots$ be events, and let

$$B = \{\omega : \omega \in A_n \text{ for an infinity of values of } n\}.$$

(i) Prove that, for any positive integer k,

$$kP(B) \leqslant \sum_{n=1}^{\infty} P(A_n).$$

Hint. Apply the result of Exercise 20 (or even Lemma 2 itself), with $X_n = I_{A_1}+I_{A_2}+ \ldots +I_{A_n}$ $(n = 1, 2, \ldots)$, and $Y = kI_B$.

(ii) Hence give an alternative proof of Theorem 2.4 (i).

22. Let X be a random variable, and let $X^{(t)} = \sin(tX)$ $(t \in R)$. Prove that

(i) $F(t) = E\{X^{(t)}\}$ is continuous on R;

(ii) $\displaystyle\int_0^t F(u)\,du = E\{Y^{(t)}\}$ $(t \in R)$,

where $Y^{(t)} = \dfrac{1}{X}\{1-\cos(tX)\}$ if $X \neq 0$, and $Y^{(t)} = 0$ if $X = 0$;

(iii) if also $E(X)$ is finite, then $F'(t)$ is finite on R and

$$F'(t) = E\{X\cos(tX)\} \quad (t \in R).$$

Note. It is implicit in the proof of Theorem 9 that $Y^{(t)}$ is a random variable with finite expected value. The reader should also give a direct proof of this.

23. Verify that the following functions are probability densities:

(i) $f(x) = 1/(b-a)$ if $a < x < b$, and $f(x) = 0$ otherwise (the *uniform distribution on* (a, b));

(ii) $f(x) = \lambda e^{-\lambda x}$ if $x > 0$, and $f(x) = 0$ otherwise, where $\lambda > 0$ (*the exponential* (λ) *distribution*);

(iii) $f(x) = \dfrac{1}{\sigma\sqrt{(2\pi)}} \exp\{-\tfrac{1}{2}(x-\mu)^2/\sigma^2\}\ (-\infty < x < \infty)$, where μ is real and $\sigma > 0$ (*the normal* (μ, σ^2) *distribution*);

(iv) $f(x) = \dfrac{1}{\pi}(1+x^2)^{-1}\ (-\infty < x < \infty)$ (the *Cauchy distribution*).

24. Let X be a random variable having one of the probability density functions of Exercise 23. Prove the following results:

(i) $E\{(X-a)^n\} = (b-a)^n/(n+1)\ (n = 1, 2, \ldots)$;

(ii) $E(X^n) = n!/\lambda^n\ (n = 1, 2, \ldots)$;

(iii) $E\{(X-\mu)^n\} = 0$ if n is odd, and $= \dfrac{(2k)!}{2^k k!}\sigma^{2k}$ if $n = 2k$ is even;

(iv) $E(X)$ does not exist.

25. Suppose that $f: R \to [0, \infty)$ is Borel measurable. Prove that the random variable X has the probability density function f if and only if

$$P(X \leqslant x) = \int_{-\infty}^{x} f(t)\, dt \quad (x \in R). \tag{$*$}$$

(Some knowledge of Lebesgue theory is needed for the solution of this exercise.)

Hint. ("if") Suppose that ($*$) holds, and let

$$Q(A) = \int_A f(t)\, dt \quad (A \in \mathscr{B}).$$

Verify that Q is a probability measure on $\mathscr{B}$, and that it has the same distribution function as X. Apply Theorem 4.2, Corollary 2, to show that $P_X = Q$.

26. Suppose that X is a random variable with distribution function F, and let n be a positive integer. Prove that $E(X^n)$ is finite if and only if the Riemann integrals

$$\int_0^{\infty} x^{n-1}\{1-F(x)\}\, dx \quad \text{and} \quad \int_{-\infty}^{0} x^{n-1}F(x)\, dx$$

are convergent, and then

$$E(X^n) = n\int_0^{\infty} x^{n-1}\{1-F(x)\}\, dx - n\int_{-\infty}^{0} x^{n-1}F(x)\, dx.$$

6 Independence

We assume throughout that a probability space $(\Omega, \mathscr{F}, P)$ is given, and that all the random variables with which we are concerned are defined on it.

Independence of two events and, more generally, of any number of events, was defined in Chapter 2. The intuitive idea which the definitions sought to make precise was that information about the occurrence of some of the events did not effect the probability of occurrence of the others.

Correspondingly, random variables are said to be independent if, roughly speaking, information about the values taken by some of them does not affect our assessment of the distribution of any of the others. Thus if X and Y are independent random variables then, for any Borel sets A and B,

$$P(X \in A \mid Y \in B) = P(X \in A)$$

(assuming $P(Y \in B) \neq 0$), or

$$P(X \in A \text{ and } Y \in B) = P(X \in A)\, P(Y \in B).$$

The second form of the condition will be preferred in the definitions which follow, because it is symmetrical in X and Y, and because its statement does not require any proviso of the form $P(Y \in B) \neq 0$. Note that the random variables X and Y will be termed independent if the events $X^{-1}(A)$ and $Y^{-1}(B)$ are independent for any choice of Borel sets A and B.

The obvious extension of this idea to any finite number of random variables yields the following definition.

Definition 1. The n random variables $X_1, X_2, \ldots, X_n$ are *independent* if the events

$$X_1^{-1}(B_1),\ X_2^{-1}(B_2),\ \ldots,\ X_n^{-1}(B_n)$$

are independent for every choice of Borel sets $B_1, B_2, \ldots, B_n$.

Thus if $X_1, X_2, \ldots, X_n$ are independent random variables then, in particular,

$$\left.\begin{array}{l} P\left\{\bigcap_{i=1}^{n} X_i^{-1}(B_i)\right\} = \prod_{i=1}^{n} P\{X_i^{-1}(B_i)\} \\ \text{for every choice of Borel sets } B_1, B_2, \ldots, B_n. \end{array}\right\} \quad (1)$$

If, conversely, the condition (1) is satisfied, then $X_1, X_2, \ldots, X_n$ are independent. This assertion is not quite trivial. We have to show that, if (1) holds, the following assertion is true:

If r is any one of the integers $2, 3, \ldots, n$,

and $i_1, i_2, \ldots, i_r$ are any r distinct integers chosen from $1, 2, \ldots, n$,

and $C_{i_1}, C_{i_2}, \ldots, C_{i_r}$ are any r Borel sets,

then

$$P\left\{\bigcap_{s=1}^{r} X_{i_s}^{-1}(C_{i_s})\right\} = \prod_{s=1}^{r} P\{X_{i_s}^{-1}(C_{i_s})\}.$$

This follows from (1) on taking $B_i = C_i$ if i is one of $i_1, i_2, \ldots, i_r$, and $B_i = R$ otherwise, and noting that, for any random variable X,

$$X^{-1}(R) = \Omega \quad \text{and} \quad P\{X^{-1}(R)\} = 1.$$

Thus an equivalent definition of independence of n random variables is the following.

Definition 1′. The n random variables $X_1, X_2, \ldots, X_n$ are *independent* if

$$P\left\{\bigcap_{i=1}^{n} X_i^{-1}(B_i)\right\} = \prod_{i=1}^{n} P\{X_i^{-1}(B_i)\}$$

for every choice of Borel sets $B_1, B_2, \ldots, B_n$.

Definition 2. The random variables of an infinite class of random variables are *independent* if the random variables of every finite sub-class are independent.

To develop the theory of independence of random variables, it will be convenient to define independence of classes of events, such classes being said to be independent if every choice of events, one from each class, yields independent events.

Definition 3 (i). Non-empty classes $\mathscr{C}_1, \mathscr{C}_2, \ldots, \mathscr{C}_n$ of sets of $\mathscr{F}$ are *independent* if $A_1, A_2, \ldots, A_n$ are independent whenever $A_i \in \mathscr{C}_i$ $(i = 1, 2, \ldots, n)$.

(ii) The classes of an infinite aggregate of non-empty classes of sets of $\mathscr{F}$ are *independent* if the classes of every finite sub-aggregate are independent.

It is an immediate consequence of the preceding definitions that the random variables X_t $(t \in T)$, where T is a non-empty index set, are independent if and only if the sigma-algebras $X_t^{-1}(\mathscr{B})$ $(t \in T)$ are independent.

In this book, those classes to which the concept of independence is applied will always be semi-algebras or, in particular, sigma-algebras. The reader will easily verify that if $\mathscr{S}_1, \mathscr{S}_2, \ldots, \mathscr{S}_n$ are semi-algebras of sets of $\mathscr{F}$, then they are independent if and only if

$$P(A_1 \cap A_2 \cap \ldots \cap A_n) = P(A_1)\,P(A_2) \ldots P(A_n)$$

whenever $A_i \in \mathscr{S}_i$ $(i = 1, 2, \ldots, n)$ (cf. the discussion preceding Definition 1′).

It can be shown that if two semi-algebras are independent, then so are the generated sigma-algebras. This result is a particular case of the following lemma, which will play an essential part in the proof of some later theorems (1(b) and 2).

Lemma 1. *Suppose that*

$$\mathscr{S}_k \quad (k \in I \cup J)$$

are independent semi-algebras of sets of $\mathscr{F}$, *where* I *and* J *are disjoint, non-empty index sets, and let*

$$\mathscr{F}_I = \sigma\left(\bigcup_{i \in I} \mathscr{S}_i\right), \quad \mathscr{F}_J = \sigma\left(\bigcup_{j \in J} \mathscr{S}_j\right).$$

Then (i) $\mathscr{S}_i$ $(i \in I)$ *and* $\mathscr{F}_J$ *are independent*;
(ii) $\mathscr{F}_I$ *and* $\mathscr{F}_J$ *are independent.*

Note. $\mathscr{F}_I$ is the smallest sigma-algebra which contains every set belonging to any of the semi-algebras $\mathscr{S}_i$ $(i \in I)$ (see Theorem 1.1). Clearly $\mathscr{F}_I \subseteq \mathscr{F}$. Corresponding statements hold for $\mathscr{F}_J$.

Proof. (i) It suffices to assume that $I = \{1, 2, \ldots, n\}$, and then to prove that

$$P(A_1 \cap A_2 \cap \ldots \cap A_n \cap B) = P(A_1)\,P(A_2) \ldots P(A_n)\,P(B)$$

whenever

$$A_i \in \mathscr{S}_i \quad (i = 1, 2, \ldots, n) \quad \text{and} \quad B \in \mathscr{F}_J. \tag{2}$$

Further, since $\mathscr{S}_1, \mathscr{S}_2, \ldots, \mathscr{S}_n$ are independent, it suffices to prove that

$$P(A_1 \cap A_2 \cap \ldots \cap A_n \cap B) = P(A_1 \cap A_2 \cap \ldots \cap A_n)\,P(B) \tag{3}$$

whenever (2) holds.

Let A be a fixed set of the form $\bigcap_{i=1}^{n} A_i$, where $A_i \in \mathscr{S}_i$ $(i = 1, 2, \ldots, n)$. Since (3) is trivial if $P(A) = 0$, we may assume that $P(A) > 0$.

Let also $\mathscr{S}$ be the class of all finite intersections of sets of $\bigcup_{j \in J} \mathscr{S}_j$, i.e. the class of all sets

$$B_1 \cap B_2 \cap \ldots \cap B_m$$

for which m is a positive integer and there exist distinct elements $j_1, j_2, \ldots, j_m$ of J such that $B_r \in \mathscr{S}_{j_r}$ $(r = 1, 2, \ldots, m)$. It is left to the reader to verify that $\mathscr{S}$ is a semi-algebra (he should note that $(B_1 \cap B_2 \cap B_3)^c$, say, is equal to

$$B_1^c \cup (B_1 \cap B_2^c) \cup (B_1 \cap B_2 \cap B_3^c),$$

and is therefore the union of a finite number of disjoint sets of $\mathscr{S}$).

For every $j \in J$

$$\mathscr{F}_J \supseteq \mathscr{S} \supseteq \mathscr{S}_j,$$

and so

$$\mathscr{F}_J \supseteq \mathscr{S} \supseteq \bigcup_{j \in J} \mathscr{S}_j.$$

Therefore

$$\mathscr{F}_J \supseteq \sigma(\mathscr{S}) \supseteq \sigma\left(\bigcup_{j \in J} \mathscr{S}_j\right) = \mathscr{F}_J,$$

and so

$$\sigma(\mathscr{S}) = \mathscr{F}_J.$$

Suppose that $B \in \mathscr{S}$. Then B can be expressed in the form $B_1 \cap B_2 \cap \ldots \cap B_m$ specified above, and

$$\begin{aligned}
P(A \cap B) &= P(A_1 \cap A_2 \cap \ldots \cap A_n \cap B_1 \cap B_2 \cap \ldots \cap B_m) \\
&= P(A_1)\,P(A_2) \ldots P(A_n)\,P(B_1)\,P(B_2) \ldots P(B_m) \\
&\quad \text{(because } \mathscr{S}_1, \mathscr{S}_2, \ldots, \mathscr{S}_n, \mathscr{S}_{j_1}, \mathscr{S}_{j_2}, \ldots, \mathscr{S}_{j_m} \\
&\quad \text{are independent)} \\
&= P(A_1 \cap A_2 \cap \ldots \cap A_n)\,P(B_1 \cap B_2 \cap \ldots \cap B_m) \\
&\quad \text{(because } \mathscr{S}_1, \mathscr{S}_2, \ldots, \mathscr{S}_n \text{ are independent} \\
&\quad \text{and } \mathscr{S}_{j_1}, \mathscr{S}_{j_2}, \ldots, \mathscr{S}_{j_m} \text{ are independent).}
\end{aligned}$$

Therefore

$$P(A \cap B) = P(A)\,P(B) \quad (B \in \mathscr{S}). \tag{4}$$

For $B \in \mathscr{F}_J$ let

$$P_1(B) = P(B \mid A) = \frac{P(A \cap B)}{P(A)}.$$

Then P_1 is a probability measure on $\mathscr{F}_J$ (see Exercise 2.4), and $P_1 = P$ on $\mathscr{S}$ (by (4)).

It now follows from the uniqueness part of Theorem 4.2 that $P_1 = P$ on $\sigma(\mathscr{S}) = \mathscr{F}_J$ i.e. that

$$\frac{P(A \cap B)}{P(A)} = P(B) \quad (B \in \mathscr{F}_J).$$

Thus (3) holds, and the proof is complete.

(ii) The second part of the lemma is established by applying the result of (i) to the independent semi-algebras $\mathscr{S}_i$ $(i \in I)$ and $\mathscr{F}_J$.

Corollary *Suppose that*

(i) $X_1, X_2, \ldots, X_m, Y_1, Y_2, \ldots, Y_n$ *are independent random variables;*
(ii) $A \in \mathscr{B}^m$, $B \in \mathscr{B}^n$.

Then

$$P(\mathbf{X} \in A \text{ and } \mathbf{Y} \in B) = P(\mathbf{X} \in A)\, P(\mathbf{Y} \in B), \tag{5}$$

where

$$\mathbf{X} = (X_1, X_2, \ldots, X_m) : \Omega \to R^m$$

and

$$\mathbf{Y} = (Y_1, Y_2, \ldots, Y_n) : \Omega \to R^n.$$

Notes. (a) $A \times B \in \mathscr{B}^{m+n}$ (by Exercise 1.10 (ii)).

(b) $\{\mathbf{X} \in A\}$, $\{\mathbf{Y} \in B\}$ and $\{\mathbf{X} \in A \text{ and } \mathbf{Y} \in B\} = \{(\mathbf{X}, \mathbf{Y}) \in A \times B\}$ are all sets of $\mathscr{F}$ (by Exercise 3.10).

(c) An m-dimensional random variable $\mathbf{X}$ and an n-dimensional random variable $\mathbf{Y}$ are said to be independent if (5) holds whenever $A \in \mathscr{B}^m$ and $B \in \mathscr{B}^n$. Thus the corollary gives a sufficient condition for two such random variables to be independent.

Proof. For $i = 1, 2, \ldots, m$ let

$$\mathscr{F}_i = X_i^{-1}(\mathscr{B}),$$

and so $\mathscr{F}_i$ is a sigma-algebra of sets of $\mathscr{F}$. Then

$$\mathscr{F}_I = \sigma(\mathscr{F}_1 \cup \mathscr{F}_2 \cup \ldots \cup \mathscr{F}_m)$$

contains all the sets

$$X_i^{-1}\{(-\infty, x]\} \qquad (i = 1, 2, \ldots, m;\ x \in R),$$

because

$$X_i^{-1}\{(-\infty, x]\} \in \mathscr{F}_i \subseteq \mathscr{F}_1 \cup \mathscr{F}_2 \cup \ldots \cup \mathscr{F}_m \subseteq \mathscr{F}_I.$$

Therefore $\mathscr{F}_I$ contains all the sets

$$\bigcap_{i=1}^{m} X_i^{-1}\{(-\infty, x_i]\} = \mathbf{X}^{-1}\{(-\infty, x_1]\times(-\infty, x_2]\times \ldots \times(-\infty, x_m]\}$$

$$(x_1, x_2, \ldots, x_m \in R).$$

Thus

$$\mathbf{X}^{-1}(\mathscr{J}^m) \subseteq \mathscr{F}_I$$

($\mathscr{J}^m$ was defined on page 6).

Therefore

$$\begin{aligned}\mathbf{X}^{-1}(\mathscr{B}^m) &= \mathbf{X}^{-1}\{\sigma(\mathscr{J}^m)\} \quad \text{(by Exercise 1.7)}\\ &= \sigma\{\mathbf{X}^{-1}(\mathscr{J}^m)\} \quad \text{(by Theorem 1.3 (ix))}\\ &\subseteq \mathscr{F}_I.\end{aligned}$$

Similarly, if

$$\mathscr{F}_j^* = Y_j^{-1}(\mathscr{B}) \qquad (j = 1, 2, \ldots, n)$$

and

$$\mathscr{F}_J = \sigma(\mathscr{F}_1^* \cup \mathscr{F}_2^* \cup \ldots \cup \mathscr{F}_n^*),$$

then

$$\mathbf{Y}^{-1}(\mathscr{B}^n) \subseteq \mathscr{F}_J.$$

Now $\mathscr{F}_1, \mathscr{F}_2, \ldots, \mathscr{F}_m, \mathscr{F}_1^*, \mathscr{F}_2^*, \ldots, \mathscr{F}_n^*$ are independent (by condition (i)), and so $\mathscr{F}_I$ and $\mathscr{F}_J$ are independent (by the lemma). The result now follows immediately because

$$\{\mathbf{X} \in A\} = \mathbf{X}^{-1}(A) \in \mathbf{X}^{-1}(\mathscr{B}^m) \subseteq \mathscr{F}_I$$

and, similarly,

$$\{\mathbf{Y} \in B\} \in \mathscr{F}_J.$$

Theorem 1 (*a*) *Suppose that*

(i) *X and Y are independent random variables;*
(ii) *$f: R \to R$ and $g: R \to R$ are Borel measurable.*

Then $f(X)$ and $g(Y)$ are independent random variables.

(*b*) *Suppose that*

(i) *$X_1, X_2, \ldots, X_m, Y_1, Y_2, \ldots, Y_n$ are independent random variables.*
(ii) *$f: R^m \to R$ and $g: R^n \to R$ are Borel measurable.*

Then $f(X_1, X_2, \ldots, X_m)$ and $g(Y_1, Y_2, \ldots, Y_n)$ are independent random variables.

Note. $f(X)$ and $g(Y)$ are random variables (by Theorem 3.5), and $f(X_1, X_2, \ldots, X_m)$ and $g(Y_1, Y_2, \ldots, Y_n)$ are random variables (by Exercise 3.11).

Proof. (a) Let A and B be any two Borel sets. Then

$$\begin{aligned}
P\{f(X) \in A \text{ and } g(Y) \in B\} \\
&= P\{X \in f^{-1}(A) \text{ and } Y \in g^{-1}(B)\} \\
&= P\{X \in f^{-1}(A)\}\, P\{Y \in g^{-1}(B)\} \quad \text{(because } X \text{ and } Y \text{ are independent, and } f^{-1}(A) \text{ and } g^{-1}(B) \in \mathscr{B}) \\
&= P\{f(X) \in A\}\, P\{g(Y) \in B\}.
\end{aligned}$$

Therefore $f(X)$ and $g(Y)$ are independent.

(b) As before, let A and B be any two Borel sets. Then $f^{-1}(A) \in \mathscr{B}^m$ and $g^{-1}(B) \in \mathscr{B}^n$, and so, by the corollary to Lemma 1,

$$\begin{aligned}
&P\{(X_1, X_2, \ldots, X_m) \in f^{-1}(A) \text{ and } (Y_1, Y_2, \ldots, Y_n) \in g^{-1}(B)\} \\
&\quad = P\{(X_1, X_2, \ldots, X_m) \in f^{-1}(A)\}\, P\{(Y_1, Y_2, \ldots, Y_n) \in g^{-1}(B)\}.
\end{aligned}$$

The rest of the proof is as in (a).

It follows from the Borel–Cantelli lemmas (Theorem 2.4) that if $A_1, A_2, \ldots$ are independent events and

$$B = \{\omega : \omega \in \text{ an infinity of } A_1, A_2, \ldots\},$$

i.e. B is the event which occurs if and only if an infinity of $A_1, A_2, \ldots$ occur, then $P(B)$ is either 0 or 1 according as $\sum_{n=1}^{\infty} P(A_n)$ is convergent or divergent. If we let $X_n = I_{A_n}$ $(n = 1, 2, \ldots)$ and $X = \overline{\lim_{n \to \infty}}\, X_n = I_B$, then the assumption that $A_1, A_2, \ldots$ are independent events is equivalent to the assumption that $X_1, X_2, \ldots$ are independent random variables, and the conclusion asserts that $P(X = 1)$ is either 0 or 1. It will be noted that the event $\{X = 1\}$ is not affected by changes in any finite number of the X_n's.

The above example is an illustration of the zero-one law. This asserts, in effect, that if $X_1, X_2, \ldots$ are independent random variables, and A is any event which depends only on the behaviour of the X_n's for all sufficiently large n, then $P(A)$ is either 0 or 1. In what follows we make this assertion precise.

Suppose that $X_1, X_2, \ldots$ are independent random variables, and let

$$\mathscr{F}_n = \sigma\left\{\bigcup_{k=n}^{\infty} X_k^{-1}(\mathscr{B})\right\} \quad (n = 1, 2, \ldots).$$

Then, for $n = 1, 2, \ldots, \mathscr{F}_n$ is the smallest sigma-algebra containing all the sets

$$X_k^{-1}(B) \quad (k = n, n+1, \ldots;\ B \in \mathscr{B}),$$

i.e. it is the smallest sigma-algebra with respect to which $X_n, X_{n+1}, \ldots$ are all measurable.

Now let
$$\mathscr{F}_\infty = \bigcap_{n=1}^{\infty} \mathscr{F}_n.$$

Then $\mathscr{F}_\infty$ is a sigma-algebra (it was shown in the proof of Theorem 1.1 that the intersection of any class of sigma-algebras is itself a sigma-algebra). Clearly
$$\mathscr{F} \supseteq \mathscr{F}_1 \supseteq \mathscr{F}_2 \supseteq \ldots \supseteq \mathscr{F}_\infty.$$

Theorem 2 (*the zero-one law*) *Suppose that* $X_1, X_2, \ldots$ *are independent random variables, and that* $\mathscr{F}_\infty$ *is defined as above. Then, for every* $A \in \mathscr{F}_\infty$, $P(A)$ *is either 0 or 1.*

Proof. Since $X_1, X_2, \ldots$ are independent random variables
$$X_1^{-1}(\mathscr{B}), X_2^{-1}(\mathscr{B}), \ldots$$
are independent sigma-algebras. Therefore, for $n = 1, 2, \ldots,$
$$X_1^{-1}(\mathscr{B}), X_2^{-1}(\mathscr{B}), \ldots, X_n^{-1}(\mathscr{B}) \quad \text{and} \quad \mathscr{F}_{n+1} = \sigma\left\{\bigcup_{k=n+1}^{\infty} X_k^{-1}(\mathscr{B})\right\}$$
are independent (by Lemma 1 (i)), and so
$$X_1^{-1}(\mathscr{B}), X_2^{-1}(\mathscr{B}), \ldots, X_n^{-1}(\mathscr{B}) \quad \text{and} \quad \mathscr{F}_\infty$$
are independent (because $\mathscr{F}_\infty \subseteq \mathscr{F}_{n+1}$).

This holds for every positive integer n, and so
$$\mathscr{F}_\infty, X_1^{-1}(\mathscr{B}), X_2^{-1}(\mathscr{B}), \ldots$$
are independent.

A further application of Lemma 1 (i) shows that
$$\mathscr{F}_\infty \quad \text{and} \quad \mathscr{F}_1 = \sigma\left\{\bigcup_{k=1}^{\infty} X_k^{-1}(\mathscr{B})\right\}$$
are independent. Therefore
$$P(A \cap B) = P(A)\,P(B)$$
whenever $A \in \mathscr{F}_\infty$ and $B \in \mathscr{F}_1$. But $\mathscr{F}_\infty \subseteq \mathscr{F}_1$ and so, in particular,
$$P(A) = P(A \cap A) = \{P(A)\}^2$$
whenever $A \in \mathscr{F}_\infty$, and the result follows.

Note. The proof shows that $\mathscr{F}_\infty$ is independent of itself (!), and is essentially indistinguishable from the sigma-algebra $\{\emptyset, \Omega\}$.

As an example of the application of the zero-one law consider independent random variables $X_1, X_2, \ldots$, and let

$$A = \{\omega\text{: the sequence } X_1(\omega), X_2(\omega), \ldots \text{ is convergent}\}.$$

It will be shown that $A \in \mathscr{F}_\infty$, and so $P(A)$ is either 0 or 1, i.e. the sequence $X_1, X_2, \ldots$ is either convergent a.s. or divergent a.s.

Now the sequence $X_1(\omega), X_2(\omega), \ldots$ is convergent if and only if to every positive integer p there corresponds a positive integer $m = m(p, \omega)$ such that

$$|X_n(\omega) - X_m(\omega)| < \frac{1}{p} \qquad (n = m+1, m+2, \ldots).$$

Therefore

$$A = \bigcap_{p=1}^{\infty} \bigcup_{m=1}^{\infty} \bigcap_{n=m+1}^{\infty} \left\{|X_n - X_m| < \frac{1}{p}\right\},$$

and so $A \in \mathscr{F}_1$. The reader should compare the above with the second proof of Theorem 3.6 (iii), and note also that though it is equally, and slightly more obviously, the case that

$$A = \bigcap_{\varepsilon > 0} \bigcup_{m=1}^{\infty} \bigcap_{n=m+1}^{\infty} \{|X_n - X_m| < \varepsilon\},$$

this does not suffice to prove that $A \in \mathscr{F}_1$.

Now the sequence $X_1(\omega), X_2(\omega), \ldots$ is convergent if and only if the sequence $X_2(\omega), X_3(\omega), \ldots$ is convergent, and so $A \in \mathscr{F}_2$, and so on. Therefore $A \in \bigcap_{n=1}^{\infty} \mathscr{F}_n = \mathscr{F}_\infty$, and so, by Theorem 2, $P(A)$ is either 0 or 1.

It can be shown further that if $P(A) = 1$ then $\lim_{n\to\infty} X_n$ is almost surely a constant. For let $X(\omega) = \lim_{n\to\infty} X_n(\omega)$ if $\omega \in A$, and $= 0$ if $\omega \notin A$. Then $X = \lim_{n\to\infty} I_A X_n$, and so X is a random variable. For any real number x let

$$A_x = \left\{\omega : \omega \in A \quad \text{and} \quad \lim_{n\to\infty} X_n(\omega) \leqslant x\right\}.$$

For $\omega \in A$, $\lim_{n\to\infty} X_n(\omega) \leqslant x$ if and only if, for every positive integer p, $X_n(\omega) < x + \frac{1}{p}$ for all sufficiently large n. Therefore

$$A_x = A \cap \bigcap_{p=1}^{\infty} \bigcup_{m=1}^{\infty} \bigcap_{n=m}^{\infty} \left\{X_n < x + \frac{1}{p}\right\} = A \cap B_x, \quad \text{say},$$

and, as before, $P(B_x)$ is either 0 or 1. Since $P(A) = 1$, $P(A_x) = P(B_x)$, and so $P(A_x)$ also is either 0 or 1.

Now, if F is the distribution function of X,

$$F(x) = P(X \leqslant x) = P(A_x),$$

because $\{X \leqslant x\} = A_x$ if $x < 0$ and $= A_x \cup A^c$ if $x \geqslant 0$, and $P(A^c) = 0$. Therefore $F(x)$ is either 0 or 1 for all $x \in R$. Consequently there exists some real number c such that $F(x) = 0$ if $x < c$ and $= 1$ if $x \geqslant c$, which implies that $P(X = c) = F(c) - F(c-) = 1$.

Theorem 3 *Suppose that*

(i) *X and Y are independent random variables;*
(ii) *$E(X)$ and $E(Y)$ are finite.*

Then $E(XY)$ is finite, and

$$E(XY) = E(X)E(Y).$$

Proof. Suppose firstly that X and Y are simple and non-negative. Let the distinct values taken by X be $a_1, a_2, \ldots, a_m$, and those taken by Y be $b_1, b_2, \ldots, b_n$. Let also

$$A_i = \{X = a_i\} \qquad (i = 1, 2, \ldots, m)$$

and

$$B_j = \{Y = b_j\} \qquad (j = 1, 2, \ldots, n).$$

Then

$$P(A_i \cap B_j) = P(A_i)P(B_j) \qquad (i = 1, 2, \ldots, m;\ j = 1, 2, \ldots, n) \tag{6}$$

(because X and Y are independent).

Now
$$\begin{aligned} E(XY) &= \sum_{i=1}^{m} \sum_{j=1}^{n} a_i b_j P(A_i \cap B_j) \quad \text{(by Definition 5.1)} \\ &= \sum_{i=1}^{m} \sum_{j=1}^{n} a_i b_j P(A_i)P(B_j) \quad \text{(by (6))} \\ &= \left\{\sum_{i=1}^{m} a_i P(A_i)\right\} \left\{\sum_{j=1}^{n} b_j P(B_j)\right\} \\ &= E(X)E(Y) \quad \text{(again by Definition 5.1).} \end{aligned}$$

Suppose secondly that X and Y are non-negative. For $n = 1, 2, \ldots$ let $g_n : R \to R$ be defined by

$$g_n(x) = \begin{cases} 0 & \text{if } x < 0 \\ (m-1)2^{-n} & \text{if } (m-1)2^{-n} \leqslant x < m2^{-n} \qquad (m = 1, 2, \ldots, n2^n) \\ n & \text{if } x \geqslant n. \end{cases}$$

Then g_n is Borel measurable (by Theorem 3.2 (ii)), and $X_n = g_n(X)$, $Y_n = g_n(Y)$ are independent random variables (by Theorem 1 (a)).

Now $X_1, X_2, \ldots, Y_1, Y_2, \ldots$ are simple random variables satisfying

$$0 \leqslant X_1 \leqslant X_2 \leqslant \ldots \text{ on } \Omega,$$

$$0 \leqslant Y_1 \leqslant Y_2 \leqslant \ldots \text{ on } \Omega,$$

$$X_n \to X \text{ on } \Omega \quad \text{as} \quad n \to \infty,$$

and

$$Y_n \to Y \text{ on } \Omega \quad \text{as} \quad n \to \infty$$

(see the proof of Theorem 3.7). Thefore $X_1Y_1, X_2Y_2, \ldots$ are simple random variables satisfying

$$0 \leqslant X_1Y_1 \leqslant X_2Y_2 \leqslant \ldots \text{ on } \Omega$$

and

$$X_nY_n \to XY \text{ on } \Omega \quad \text{as} \quad n \to \infty.$$

Therefore
$$\begin{aligned} E(XY) &= \lim_{n\to\infty} E(X_nY_n) && \text{(by Definition 5.2)} \\ &= \lim_{n\to\infty} E(X_n)E(Y_n) && \text{(by what has just been proved)} \\ &= E(X)E(Y) && \text{(again by Definition 5.2).} \end{aligned}$$

(and so, in particular, $E(XY)$ is finite).

Suppose finally that X and Y can take values of either sign. With the notation of Chapter 5,

$$X = X^+ - X^-, \quad Y = Y^+ - Y^-,$$

and $E(X^+)$, $E(X^-)$, $E(Y^+)$, $E(Y^-)$ are all finite.

Let $f : R \to R$ and $g : R \to R$ be defined by

$$f(x) = \tfrac{1}{2}(|x|+x) \quad (x \in R)$$

and

$$g(x) = \tfrac{1}{2}(|x|-x) \quad (x \in R).$$

Then, as before, $X^+ = f(X)$ and $Y^- = g(Y)$, for example, are independent, and so, by what has already been proved, $E(X^+Y^-) = E(X^+)E(Y^-)$.

Therefore, by Theorems 5.1 and 5.2, $E(XY)$ is finite and

$$\begin{aligned} E(XY) &= E(X^+Y^+) - E(X^+Y^-) - E(X^-Y^+) + E(X^-Y^-) \\ &= E(X^+)E(Y^+) - E(X^+)E(Y^-) - E(X^-)E(Y^+) + E(X^-)E(Y^-) \\ &= \{E(X^+) - E(X^-)\}\{E(Y^+) - E(Y^-)\} \\ &= E(X)E(Y) \quad \text{(by Definition 5.3).} \end{aligned}$$

Corollary *Suppose that*

(i) $X_1, X_2, \ldots, X_n$ *are independent random variables*;
(ii) $E(X_i)$ *is finite* $(i = 1, 2, \ldots, n)$.

Then $E(X_1X_2 \dots X_n)$ is finite and

$$E(X_1X_2 \dots X_n) = E(X_1)E(X_2) \dots E(X_n).$$

First proof. This proceeds by induction over n.

Let the statement of the corollary be denoted by $S(n)$. Then $S(2)$ is true (by the theorem). Now suppose that $S(n)$ is true for some $n \geqslant 2$ (the inductive hypothesis), and that $X_1, X_2, \dots, X_{n+1}$ are independent random variables with finite expectations. Then $X_1X_2 \dots X_n$ and X_{n+1} are independent (by Theorem 1(b)), and $E(X_1X_2 \dots X_n)$ is finite (by the inductive hypothesis). Therefore, by the theorem, $E(X_1X_2 \dots X_nX_{n+1})$ is finite, and

$$\begin{aligned} E(X_1X_2 \dots X_nX_{n+1}) &= E(X_1X_2 \dots X_n)E(X_{n+1}) \\ &= E(X_1)E(X_2) \dots E(X_n)E(X_{n+1}) \end{aligned}$$

(by the inductive hypothesis).

Therefore $S(n+1)$ is true if $S(n)$ is, and the proof is complete.

The reader should note the part played by Theorem 1(b) in this proof.

Second proof. This follows the pattern of the proof of the theorem itself, the result being established first for random variables which are simple and non-negative, then for those which are non-negative, and finally for those which can take values of either sign.

The details, which are perfectly straightforward, are left to the reader.

The essential part played by the independence condition in the above theorem will be brought home to the reader if he considers the following examples, in each of which

$$\Omega = \{1, 2, \dots\}$$

and $$P(\{n\}) = \frac{1}{n(n+1)} \qquad (n = 1, 2, \dots).$$

(a) $X(n) = n, \quad Y(n) = 0 \quad \text{if} \quad n$ is odd

and $X(n) = 0, \quad Y(n) = n \quad \text{if} \quad n$ is even.

Then $E(X) = E(Y) = \infty$ and $E(XY) = 0$.

(b) $X(n) = Y(n) = n^{\frac{1}{2}}$ (all n).

Then $E(X), E(Y)$ are finite, and $E(XY) = \infty$.

(c) $X(1) = 1, \quad Y(1) = 0$

and $X(n) = 0, \quad Y(n) = 1 \quad \text{if} \quad n > 1$.

Then $E(X) = E(Y) = \frac{1}{2}$ and $E(XY) = 0$.

Finally we give an example to show that if $E(X)$, $E(Y)$ are finite and $E(XY) = E(X)E(Y)$, it does not follow that X and Y are independent.

(d) $X(1) = 2, \quad Y(1) = 1, \quad X(2) = 0, \quad Y(2) = 3$

and $X(n) = Y(n) = 0$ if $n > 2$.

Then $E(X) = E(Y) = E(XY) = 1$.

In each case the reader will easily verify that X and Y are not independent.

Exercises

1. Let $\mathscr{C}$ be a class of independent random variables. Prove that the random variables of any sub-class of $\mathscr{C}$ are independent.

2. Let T be a non-empty index set, and let A_t $(t \in T)$ be events.

(i) Prove that the events A_t $(t \in T)$ are independent if and only if the sigma-algebras $\{\emptyset, A_t, A_t^c, \Omega\}$ $(t \in T)$ are independent.

(ii) Prove that the events A_t $(t \in T)$ are independent if and only if their indicator functions I_{A_t} $(t \in T)$ are independent.

3. Show that the random variable X is independent of itself if and only if there exists a real number c such that $X = c$ a.s.

Hint. ("only if") See Exercise 4.8.

4. Let X and Y be independent random variables for which $E(X+Y)$ is finite. Prove that $E(X)$ and $E(Y)$ are finite.

Hint. There exists a positive integer k such that $p = P(|Y| \leqslant k) > 0$. Show that

$$pP(|X| \geqslant n+k) \leqslant P(|X+Y| \geqslant n) \qquad (n = 1, 2, \ldots).$$

Now use Exercise 5.18.

5. Suppose that $X_1, X_2, \ldots, X_n$ are independent random variables, and that X_i has the distribution function F_i $(i = 1, 2, \ldots, n)$. What are the distribution functions of $\max(X_1, X_2, \ldots, X_n)$ and $\min(X_1, X_2, \ldots, X_n)$?

6. The random variables $X_1, X_2, \ldots$ are independent with common distribution function F, and $Y_n = \max(X_1, X_2, \ldots, X_n)$ $(n = 1, 2, \ldots)$. Prove the following results:

(i) If $x^a\{1-F(x)\} \to b$ as $x \to \infty$, where $a > 0$ and $b > 0$, then, as $n \to \infty$, $P\{Y_n \leqslant x(bn)^{1/a}\} \to \exp(-x^{-a})$ if $x > 0$, and $\to 0$ if $x \leqslant 0$.

(ii) If $e^x\{1-F(x)\} \to b$ as $x \to \infty$, where $b > 0$, then, as $n \to \infty$, $P\{Y_n \leqslant x+\log(nb)\} \to \exp(-e^{-x})$ for all real x.

7. (i) Suppose that $X_1, X_2, \ldots, X_n$ are independent random variables, and that $f_i : R \to R$ is Borel measurable ($i = 1, 2, \ldots, n$). Prove that $f_1(X_1), f_2(X_2), \ldots, f_n(X_n)$ are independent.

(ii) State and prove the result that bears the same relation to that of (i) as Theorem 1(b) does to Theorem 1(a).

8. Suppose that the random variables $X_1, X_2, \ldots$ are independent, and let

$$A = \left\{\omega : \sum_{n=1}^{\infty} X_n(\omega) \text{ is convergent}\right\}.$$

Prove that $P(A)$ is either 0 or 1.

9. Let the random variables $X_0, X_1, X_2, \ldots$ be independent. Prove that, for any $x \in (0, \infty)$, the series $\sum_{n=0}^{\infty} X_n x^n$ is either convergent a.s. or divergent a.s.

Hence show that the radius of convergence of the series is a.s. R for some $R \in [0, \infty]$ in the sense that, with probability one, the series is absolutely convergent for all x for which $|x| < R$ and is divergent for all x for which $|x| > R$.

Hint. Consider the sets

$$A_r = \left\{\omega : \text{the series } \sum_{n=0}^{\infty} X_n(\omega) r^n \text{ is convergent}\right\} \quad (r > 0),$$

and define R to be $\inf\{r : P(A_r) = 0\}$, with an obvious interpretation if $\{r : P(A_r) = 0\} = \emptyset$. If, say, $0 < R < \infty$, investigate the behaviour of $\sum_{n=0}^{\infty} X_n(\omega) x^n$ for $\omega \in \bigcap_{k=1}^{\infty} (A_{R-1/k} \cap A^c_{R+1/k})$ (here A_r is to be taken as Ω if $r \leqslant 0$).

Note. If also $X_0, X_1, X_2, \ldots$ are identically distributed, it can be shown that R must take one of the values 0, 1 or ∞.

10. Suppose that $X_1, X_2, \ldots, X_n$ are independent random variables with finite second moments. Prove that

$$\operatorname{var}\left(\sum_{i=1}^{n} X_i\right) = \sum_{i=1}^{n} \operatorname{var} X_i.$$

11. The random variables X and Y are independent and have finite second moments. Prove that

$$\operatorname{var}(XY) = (\operatorname{var} X)(\operatorname{var} Y) + \{E(X)\}^2 (\operatorname{var} Y) + \{E(Y)\}^2 (\operatorname{var} X).$$

12. Suppose that X and Y are independent random variables, and that $E(X)$, $E(Y)$ and $E\{(X+Y)^2\}$ are finite. Prove that $E(X^2)$ and $E(Y^2)$ are finite.

13. Suppose that $X_1, X_2, \ldots, X_n$ are independent random variables, and let $F_1, F_2, \ldots, F_n$ and F be the distribution functions of $X_1, X_2, \ldots, X_n$ and $\mathbf{X} = (X_1, X_2, \ldots, X_n)$ respectively (see Exercise 4.12). Prove that $X_1, X_2, \ldots, X_n$ are independent if and only if

$$F(x_1, x_2, \ldots, x_n) = F_1(x_1)\, F_2(x_2) \ldots F_n(x_n)$$

for all real numbers $x_1, x_2, \ldots, x_n$.

Hint. The "only if" part is trivial. To prove the "if" part, consider the semi-algebras $\mathscr{S}_i = X_i^{-1}(\mathscr{I})$ $(i = 1, 2, \ldots, n)$. The given condition ensures that $\mathscr{S}_1, \mathscr{S}_2, \ldots, \mathscr{S}_n$ are independent, and therefore, by repeated applications of Lemma 1, that $\sigma(\mathscr{S}_1), \sigma(\mathscr{S}_2), \ldots, \sigma(\mathscr{S}_n)$ are independent. Since

$$\sigma(\mathscr{S}_i) = \sigma\{X_i^{-1}(\mathscr{I})\} = X_i^{-1}\{\sigma(\mathscr{I})\} = X_i^{-1}(\mathscr{B}) \qquad (i = 1, 2, \ldots, n),$$

it follows that $X_1, X_2, \ldots, X_n$ are independent.

The following two exercises outline proofs of the weak law of large numbers (Exercise 14) and the strong law of large numbers (Exercise 15), in each case under a restrictive condition. The general statements and proofs of these two laws for independent and identically distributed random variables will be given later (see Theorems 8.6 and D.1).

14. Suppose that $X_1, X_2, \ldots$ are independent and identically distributed random variables with finite second moment. Let $\mu = E(X_1)$, and

$$S_n = X_1 + X_2 + \ldots + X_n \qquad (n = 1, 2, \ldots).$$

Prove that, for every $\varepsilon > 0$,

$$P\left(\left|\frac{1}{n}S_n - \mu\right| \geqslant \varepsilon\right) \to 0 \quad \text{as} \quad n \to \infty.$$

Hint. Apply Chebyshev's inequality (Exercise 5.8) to obtain

$$P(|S_n - n\mu| \geqslant n\varepsilon) \leqslant (\operatorname{var} X_1)/(n\varepsilon^2).$$

15. Suppose that $X_1, X_2, \ldots$ are independent and identically distributed random variables with finite fourth moment. Let $\mu = E(X_1)$, and

$$S_n = X_1 + X_2 + \ldots + X_n \qquad (n = 1, 2, \ldots).$$

Prove that
$$P\left(\frac{1}{n}S_n \to \mu \quad \text{as} \quad n \to \infty\right) = 1,$$

i.e. that
$$\frac{1}{n}S_n \to \mu \quad \text{a.s. as} \quad n \to \infty.$$

Outline of solution. There is no loss of generality in assuming that $\mu = 0$.

(i) Prove that $E(S_n^4) = n\mu_4 + 3n(n-1)\mu_2^2$, where $\mu_2 = E(X_1^2)$, $\mu_4 = E(X_1^4)$.

(ii) Prove that, for every $\varepsilon > 0$,

$$p_n(\varepsilon) = P\left(\frac{1}{n}|S_n| \geqslant \varepsilon\right) \leqslant \{n\mu_4 + 3n(n-1)\mu_2^2\}/(n^4\varepsilon^4).$$

Hint. Use Exercise 5.10 (i) with $f(x) = x^4$.

(iii) Deduce that $\sum_{n=1}^{\infty} p_n(\varepsilon)$ is convergent ($\varepsilon > 0$).

(iv) Let $A(\varepsilon) = \left\{\omega : \frac{1}{n}|S_n(\omega)| \geqslant \varepsilon \text{ for an infinity of values of } n\right\}$.

Use the Borel–Cantelli lemmas to prove that $P\{A(\varepsilon)\} = 0$.

(v) Let $A = \bigcup_{k=1}^{\infty} A\left(\frac{1}{k}\right)$. Prove that $P(A) = 0$.

(vi) Observe that if $\omega \in A^c$ then, for every positive integer k,

$$\frac{1}{n}|S_n(\omega)| < \frac{1}{k}$$

for all sufficiently large values of n, and so

$$\frac{1}{n}S_n(\omega) \to 0 \quad \text{as} \quad n \to \infty.$$

7
Convergence

Let X_1, X_2, ... and X be random variables. The aim of this chapter is to define those modes of convergence of the sequence $\{X_n\}$ to the limit X which are of special interest in probability theory, to discuss their interrelations, and to give proofs of some of the fundamental results concerning them; in particular, those we shall need later in this book. We shall not aim at anything approaching a complete treatment, but rather at laying a sound foundation on which, we hope, the reader will find it easy to build by reference to more advanced texts.

The most familiar modes of convergence are those denoted by

$$X_n \to X \text{ on } \Omega \quad \text{as} \quad n \to \infty$$

and

$$X_n \to X \text{ uniformly on } \Omega \quad \text{as} \quad n \to \infty.$$

The first means that to each $\varepsilon > 0$ and each $\omega \in \Omega$ corresponds a positive integer $m = m(\varepsilon, \omega)$ such that

$$|X_n(\omega) - X(\omega)| < \varepsilon \quad (n > m).$$

The second means that to each $\varepsilon > 0$ corresponds a positive integer $m = m(\varepsilon)$ such that

$$|X_n(\omega) - X(\omega)| < \varepsilon \quad (n > m;\ \omega \in \Omega).$$

However, although these two modes of convergence are familiar, and of importance in analysis generally, there are others which are of special interest in probability theory, and these we proceed to introduce.

We shall assume throughout this chapter that, unless otherwise stated, X_1, X_2, ... and X are random variables defined on a probability space $(\Omega, \mathscr{F}, P)$, with corresponding distribution functions F_1, F_2, ... and F.

Definition 1. The sequence $\{X_n\}$ is *almost surely convergent* to X ($X_n \to X$ a.s. as $n \to \infty$) if there exists a set A of $\mathscr{F}$ such that $P(A) = 0$ and

$$X_n \to X \text{ on } A^c \quad \text{as} \quad n \to \infty$$

(see page 86).

Clearly, if $X_n \to X$ on Ω as $n \to \infty$, or if $X_n \to X$ uniformly on Ω as $n \to \infty$, then $X_n \to X$ a.s. as $n \to \infty$. The converse statements are clearly false.

We shall meet almost sure convergence later in connection with the strong law of large numbers.

For our next definition we shall assume that

(i) r is a positive real number;
(ii) $E(|X_n|^r) < \infty$ $(n = 1, 2, \ldots)$ and $E(|X|^r) < \infty$.

It is a simple exercise to show that

$$(1+t)^r \leqslant c_r(1+t^r) \quad (t \geqslant 0),$$

where $c_r = 1$ if $r \leqslant 1$, and $= 2^{r-1}$ if $r \geqslant 1$. It follows that

$$|x+y|^r \leqslant c_r(|x|^r+|y|^r)$$

for all real x and y, and hence that

$$E(|X_n-X|^r) < \infty \qquad (n = 1, 2, \ldots).$$

Definition 2. The sequence $\{X_n\}$ *converges* to X *in rth mean* ($X_n \xrightarrow{r} X$ as $n \to \infty$) if

$$E(|X_n-X|^r) \to 0 \quad \text{as} \quad n \to \infty.$$

It is left as an exercise to show that if $X_n \to X$ uniformly on Ω as $n \to \infty$ and condition (ii) above holds then $X_n \xrightarrow{r} X$ as $n \to \infty$.

However, if $X_n \to X$ a.s. as $n \to \infty$ it does not follow that $X_n \to X$ as $n \to \infty$.

Example 1. Let $\Omega = \{1, 2, \ldots\}$ and $\mathscr{F} = \mathscr{P}(\Omega)$, and let P be defined by $P(\{n\}) = \dfrac{1}{n(n+1)}$ $(n = 1, 2, \ldots)$. Let X_n be defined by $X_n(\omega) = 2^n$ if $\omega = n$, and $= 0$ if $\omega \neq n$ $(n = 1, 2, \ldots)$, and let $X(\omega) = 0$ $(\omega \in \Omega)$. Then

$$X_n \to X \text{ on } \Omega \quad \text{as} \quad n \to \infty$$

and condition (ii) holds for each $r > 0$, but

$$E(|X_n-X|^r) \to \infty \quad \text{as} \quad n \to \infty \quad (r > 0).$$

Nor, if $X_n \xrightarrow{r} X$ as $n \to \infty$ for some $r > 0$ (or, indeed, for all $r > 0$) does it follow that $X_n \to X$ a.s. as $n \to \infty$.

Example 2. Let $\Omega = (0, 1]$, $\mathscr{F} =$ the class of those Borel sets contained in $(0, 1]$ (i.e. $\mathscr{F} = \mathscr{B} \cap \mathscr{P}(\Omega)$), and $P =$ Lebesgue measure on $\mathscr{F}$. Then if

$$X_1, X_2, X_3, X_4, X_5, X_6, X_7, \ldots$$

are the indicator functions of

$$(0, \tfrac{1}{2}], (\tfrac{1}{2}, 1], (0, \tfrac{1}{3}], (\tfrac{1}{3}, \tfrac{2}{3}], (\tfrac{2}{3}, 1], (0, \tfrac{1}{4}], (\tfrac{1}{4}, \tfrac{1}{2}], \ldots$$

respectively, and $X = 0$ on Ω, it is easy to see that

$$E(|X_n - X|^r) \to 0 \quad \text{as} \quad n \to \infty \quad (r > 0),$$

but

$$X_n(\omega) \not\to X(\omega) \quad \text{as} \quad n \to \infty \quad (\omega \in \Omega).$$

Suppose that $r > 0$, and let L_r be the class of those random variables X which satisfy

$$E(|X|^r) < \infty.$$

If two random variables which are equal a.s. are regarded as identical, then L_r becomes a metric space if the distance $d(X, Y)$ between two random variables X, Y of L_r is defined to be

$$\begin{cases} E(|X-Y|^r) & \text{if} \quad 0 < r \leqslant 1 \\ \{E(|X-Y|^r)\}^{1/r} & \text{if} \quad r \geqslant 1. \end{cases}$$

It can be shown that if $X_1, X_2, \ldots \in L_r$, and

$$E(|X_m - X_n|^r) \to 0 \quad \text{as} \quad m, n \to \infty,$$

then there exists a random variable $X \in L_r$ such that

$$X_n \xrightarrow{r} X \quad \text{as} \quad n \to \infty$$

(compare Exercise 2 (i)), and so L_r is complete. It can be shown further that L_r is a Banach space if $r \geqslant 1$, and is a Hilbert space if $r = 2$.

The most important values of r from the point of view of probability theory are $r = 1$ and $r = 2$, when it is customary to speak of *convergence in mean* and *convergence in mean square* respectively.

Definition 3. The sequence $\{X_n\}$ *converges* to X *in probability* ($X_n \xrightarrow{P} X$ as $n \to \infty$) if

$$P(|X_n - X| \geqslant \varepsilon) \to 0 \quad \text{as} \quad n \to \infty$$

for every $\varepsilon > 0$.

Theorem 1 *Suppose that $X_1, X_2, \ldots, X$ are random variables, and that*

either (i) $X_n \to X$ *a.s. as* $n \to \infty$,

or (ii) $X_n \xrightarrow{r} X$ *as* $n \to \infty$ *for some* $r > 0$.

Then
$$X_n \xrightarrow{P} X \quad \textit{as} \quad n \to \infty.$$

Proof. Suppose that $\varepsilon > 0$.

(i) There exists a set A of $\mathscr{F}$ such that $P(A) = 0$ and $X_n \to X$ on A^c as $n \to \infty$.

Let $\quad B_n = \{|X_m - X| < \varepsilon \quad (m \geqslant n)\} \quad (\in \mathscr{F}) \qquad (n = 1, 2, \ldots).$

Then
$$B_1 \subseteq B_2 \subseteq \ldots \quad \text{and} \quad B = \bigcup_{n=1}^{\infty} B_n \supseteq A^c.$$

Since $P(A^c) = 1$, $P(B) = 1$, and so $P(B_n) \to 1$ as $n \to \infty$ (by Theorem 2.2).

Therefore, since $\{|X_n - X| \geqslant \varepsilon\} \subseteq B_n^c \qquad (n = 1, 2, \ldots)$,
$$\begin{aligned} P(|X_n - X| \geqslant \varepsilon) &\leqslant P(B_n^c) \\ &\to 0 \quad \text{as} \quad n \to \infty. \end{aligned}$$

(ii) Let $A_n = \{|X_n - X| \geqslant \varepsilon\} \qquad (n = 1, 2, \ldots)$.

Since
$$|X_n - X|^r \geqslant |X_n - X|^r I_{A_n} \geqslant \varepsilon^r I_{A_n} \qquad (n = 1, 2, \ldots),$$
it follows that
$$\begin{aligned} P(A_n) &\leqslant \varepsilon^{-r} E(|X_n - X|^r) \\ &\to 0 \quad \text{as} \quad n \to \infty. \end{aligned}$$

(In effect, we have used the result of Exercise 5.10 (i), with $f(x) = x^r$).

If $X_n \xrightarrow{P} X$ as $n \to \infty$, it does not follow that $X_n \to X$ a.s. as $n \to \infty$ (see Example 2, above) or that $X_n \xrightarrow{r} X$ as $n \to \infty$ for any $r > 0$ (see Example 1, above).

It can be shown that if $X_1, X_2, \ldots$ are random variables for which
$$P(|X_m - X_n| \geqslant \varepsilon) \to 0 \quad \text{as} \quad m, n \to \infty$$
for every $\varepsilon > 0$, then there exists a random variable X such that $X_n \xrightarrow{P} X$ as $n \to \infty$ (compare Exercise 2 (ii)).

We shall meet convergence in probability later in connection with the weak law of large numbers.

Definition 4. The sequence $\{X_n\}$ *converges* to X *in distribution* ($X_n \xrightarrow{D} X$ as $n \to \infty$) if
$$F_n(x) \to F(x) \quad \text{as} \quad n \to \infty \tag{1}$$
whenever x is a point of continuity of F.

Note. In Definition 4 the random variables play no direct part, appearing only in the form of their distribution functions. Given distribution functions $F_1, F_2, \ldots$ and F, it will, on occasion, be convenient to omit any reference to underlying random variables and let

$$F_n \xrightarrow{D} F \quad \text{as} \quad n \to \infty$$

mean that (1) holds whenever x is a point of continuity of F.

That the condition (1) is required to hold if F is continuous at x, but not necessarily otherwise, may at first sight seem a little surprising. The reasonableness of some such condition may, however, be seen by taking X_n to be $\frac{1}{n}$ on Ω ($n = 1, 2, \ldots$), and X to be 0 on Ω. For any acceptable meaning of the word "converges", the sequence $\{X_n\}$ converges to X. However

$$F_n(x) = \begin{cases} 1 & \text{if} \quad x \geqslant \frac{1}{n} \\ 0 & \text{if} \quad x < \frac{1}{n} \end{cases} \qquad \text{(all } n\text{)},$$

and

$$F(x) = \begin{cases} 1 & \text{if} \quad x \geqslant 0 \\ 0 & \text{if} \quad x < 0. \end{cases}$$

Although $F_n(x) \to F(x)$ as $n \to \infty$ for all $x \neq 0$, it is not true that

$$F_n(0)\,(= 0) \to F(0)\,(= 1) \quad \text{as} \quad n \to \infty.$$

Theorem 2 *Suppose that* $X_n \xrightarrow{P} X$ *as* $n \to \infty$. *Then* $X_n \xrightarrow{D} X$ *as* $n \to \infty$.

Proof. Suppose that F is continuous at x. Given $\varepsilon > 0$, choose u, v so that $u < x < v$ and

$$F(u) > F(x) - \varepsilon, \quad F(v) < F(x) + \varepsilon \tag{2}$$

(this can be done, since F is continuous at x). Then

$$\begin{aligned} F_n(x) &= P(X_n \leqslant x) \\ &= P(X_n \leqslant x \text{ and } X \leqslant v) + P(X_n \leqslant x \text{ and } X > v) \\ &\leqslant P(X \leqslant v) + P(|X_n - X| \geqslant v - x) \\ &\to P(X \leqslant v) = F(v) \quad \text{as} \quad n \to \infty. \end{aligned}$$

Similarly

$$P(X \leqslant u) \leqslant P(X_n \leqslant x) + P(|X_n - X| \geqslant x - u),$$

and so

$$\begin{aligned} F_n(x) &\geqslant P(X \leqslant u) - P(|X_n - X| \geqslant x - u) \\ &\to F(u) \quad \text{as} \quad n \to \infty. \end{aligned}$$

Therefore, by (2),

$$F(x)-\varepsilon < F_n(x) < F(x)+\varepsilon$$

for all sufficiently large n, and so

$$F_n(x) \to F(x) \quad \text{as} \quad n \to \infty.$$

However, if $X_n \xrightarrow{D} X$ as $n \to \infty$, it does not follow that $X_n \xrightarrow{P} X$ as $n \to \infty$. For let $\Omega = \{1, 2\}$ and $\mathscr{F} = \mathscr{P}(\Omega)$, and let P be defined by $P(\{1\}) = P(\{2\}) = \frac{1}{2}$. Let $X_n(1) = 1$, $X_n(2) = 0$ (all n), and $X(1) = 0$, $X(2) = 1$. Then

$$F_n(x) = F(x) \quad (x \in R;\ \text{all } n),$$

but

$$P(|X_n - X| \geqslant \tfrac{1}{2}) = 1 \quad (\text{all } n).$$

The interrelations between the various modes of convergence may be summarised in the following diagram, which shows all the possible implications.

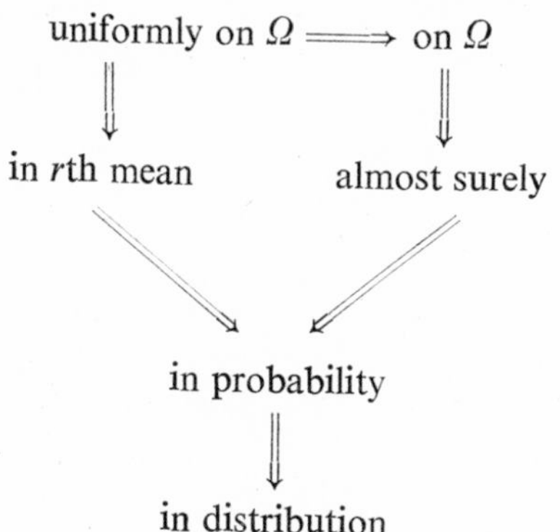

As shown above, the converse of Theorem 2 is false. The next theorem shows that the converse holds if we make the additional assumption that X is constant on Ω.

Theorem 3 *Suppose that $X_n \xrightarrow{D} c$ as $n \to \infty$ (i.e. $X_n \xrightarrow{D} X$ as $n \to \infty$, where $X(\omega) = c\ (\omega \in \Omega)$). Then $X_n \xrightarrow{P} c$ as $n \to \infty$.*

Proof. Let F be the distribution function of the constant random variable c. Then $F(x) = 0$ if $x < c$, and $= 1$ if $x \geqslant c$, and so F is continuous at all points other than c.

Now suppose that $\varepsilon > 0$. Then

$$\begin{aligned} P(|X_n - c| \geqslant \varepsilon) &= P(X_n \leqslant c-\varepsilon)+P(X_n \geqslant c+\varepsilon) \\ &\leqslant P(X_n \leqslant c-\varepsilon)+P(X_n > c+\tfrac{1}{2}\varepsilon) \\ &= F_n(c-\varepsilon)+\{1-F_n(c+\tfrac{1}{2}\varepsilon)\} \\ &\to F(c-\varepsilon)+\{1-F(c+\tfrac{1}{2}\varepsilon)\} \quad \text{as} \quad n \to \infty \end{aligned}$$

(by Definition 4). The result follows, since

$$F(c-\varepsilon)+\{1-F(c+\tfrac{1}{2}\varepsilon)\} = 0+\{1-1\} = 0.$$

To draw useful conclusions from the assertion that $X_n \xrightarrow{D} X$ as $n \to \infty$, we have to know that there are not too many points at which the distribution function F of X is discontinuous. That information is provided by the corollary to the following lemma.

Lemma 1. *Suppose that the function F is non-decreasing and bounded on R, and let D be the set of those points x of R at which F is not continuous. Then D is countable.*

Proof. There is no loss of generality in assuming that $0 \leqslant F \leqslant 1$ on R.

For all $x \in R$, $F(x-)$ and $F(x+)$ exist, and $F(x-) \leqslant F(x) \leqslant F(x+)$. Also $x \in D$ if and only if $F(x+) > F(x-)$.

For $n = 1, 2, \ldots$ let

$$A_n = \left\{x : F(x+)-F(x-) > \frac{1}{n}\right\}.$$

Then $D = \bigcup_{n=1}^{\infty} A_n$.

It will be shown that, for each n, A_n contains fewer than n points. Thus every A_n is finite, and so D is countable.

Suppose that, for some n, A_n contains at least n points. Then there exist real numbers $x_1, x_2, \ldots, x_n$ of A_n satisfying $x_1 < x_2 < \ldots < x_n$. Therefore

$$0 \leqslant F(x_1-) < F(x_1+) \leqslant F(x_2-) < F(x_2+) \leqslant \ldots \leqslant F(x_n-) < F(x_n+) \leqslant 1,$$

and so

$$\sum_{k=1}^{n} \{F(x_k+)-F(x_k-)\} \leqslant 1.$$

But, by the definition of A_n,

$$\sum_{k=1}^{n} \{F(x_k+)-F(x_k-)\} > n \times \frac{1}{n} = 1.$$

This contradiction completes the proof of the lemma.

Note. It is not, in fact, necessary to assume that F is bounded for the conclusion of the lemma to hold. The extension of the lemma to unbounded F (which will not be needed in this book) is left to the reader.

Corollary *Suppose that F is a distribution function. Then any interval of the real line contains a non- countable infinity of points at which F is continuous (and so, in particular, at least one such point).*

Proof. Any interval of the real line contains a non-countable infinity of points.

The following theorem gives a characterisation of convergence in distribution.

Theorem 4 *Suppose that $X_1, X_2, \ldots$ and X are random variables. Then*

$$X_n \xrightarrow{D} X \quad as \quad n \to \infty$$

if and only if

$$E\{f(X_n)\} \to E\{f(X)\} \quad as \quad n \to \infty$$

for every function $f: R \to R$ which is bounded and continuous on R.

Note. $E\{f(X_n)\}$ $(n = 1, 2, \ldots)$ and $E\{f(X)\}$ are finite (by Theorem 5.4, corollary).

Proof. We content ourselves with giving the proof of the "only if" part of the theorem. The "if" part of the theorem will not be needed in this book, and its proof is left as an exercise (Exercise 17).

As convergence in distribution involves the distribution functions of the random variables concerned, it will be convenient to express the expected values by means of Riemann–Stieltjes integrals (see Theorem 5.11).

Suppose that $\varepsilon > 0$, and choose real numbers a, b with $a < b$ such that F is continuous at a and b, and

$$F(a) < \varepsilon, \quad F(b) > 1-\varepsilon. \tag{3}$$

(That this can be done follows from Theorem 4.1 (iii) and Lemma 1, corollary.)

Then
$$\begin{aligned} E\{f(X)\} &= \int_{-\infty}^{\infty} f\,dF \\ &= \int_{-\infty}^{a} f\,dF + \int_{a}^{b} f\,dF + \int_{b}^{\infty} f\,dF. \end{aligned}$$

Let $M = \sup\{|f(x)| : x \in R\}$. Then

$$\left|\int_{-\infty}^{a} f\,dF\right| \leqslant \int_{-\infty}^{a} |f|\,dF \leqslant MF(a) \leqslant M\varepsilon,$$

and, similarly, $\left|\int_{b}^{\infty} f\,dF\right| \leqslant M\varepsilon$. Therefore

$$\left|E\{f(X)\} - \int_{a}^{b} f\,dF\right| \leqslant 2M\varepsilon. \tag{4}$$

Since a and b are points of continuity of F, and $X_n \xrightarrow{D} X$ as $n \to \infty$,

$$F_n(a) \to F(a) \quad \text{and} \quad F_n(b) \to F(b) \quad \text{as} \quad n \to \infty.$$

Therefore, by (3),

$$F_n(a) < \varepsilon, \quad F_n(b) > 1-\varepsilon$$

for all sufficiently large n, for $n > n_0$, say. Therefore, as before

$$\left| E\{f(X_n)\} - \int_a^b f\,dF_n \right| \leqslant 2M\varepsilon \quad (n > n_0). \tag{5}$$

Therefore, by (4) and (5),

$$|E\{f(X_n)\} - E\{f(X)\}| \leqslant \left| \int_a^b f\,dF_n - \int_a^b f\,dF \right| + 4M\varepsilon \quad (n > n_0). \tag{6}$$

By Theorem C.8

$$\int_a^b f\,dF_n \to \int_a^b f\,dF \quad \text{as} \quad n \to \infty.$$

Therefore, by (6),

$$|E\{f(X_n)\} - E\{f(X)\}| < \varepsilon(4M+1)$$

for all sufficiently large n.

This completes the proof.

The laws of large numbers

Suppose that $X_1, X_2, \ldots$ are independent and identically distributed random variables with common finite expected value μ, and let

$$\bar{X}_n = \frac{1}{n}(X_1 + X_2 + \ldots + X_n) \qquad (n = 1, 2, \ldots).$$

Then it can be shown that

$$\bar{X}_n \xrightarrow{P} \mu \quad \text{as} \quad n \to \infty \tag{7}$$

(the *weak law of large numbers*), and that

$$\bar{X}_n \to \mu \text{ a.s.} \quad \text{as} \quad n \to \infty \tag{8}$$

(the *strong law of large numbers*).

If we assume additionally that $X_1, X_2, \ldots$ have a finite second moment, (7) is an immediate consequence of Chebyshev's inequality (see Exercise 6.14). To give a general proof of (7) with no additional assumptions it suffices, once the strong law has been established, to note that the weak law then follows by Theorem 1 (i). A general proof of the weak law which does not require a prior proof of the strong law will be given in the next chapter (see Theorem 8.6), where it will appear as a consequence of certain results on characteristic functions which we shall establish there.

If we assume additionally that $X_1, X_2, \ldots$ have a finite fourth moment, it is possible to give a relatively simple proof of (8); this was outlined in Exercise 6.15. For a general proof see Theorem D.1.

Although we postpone the general proofs of the two laws, it will be convenient to discuss them in this chapter, and that we now proceed to do, paying particular attention to their significance.

The initial axioms of probability theory were suggested by the long-term behaviour of relative frequencies. If the theory is a good one, we would expect, among other things, that such long-term behaviour (appropriately defined) would be a consequence of the axioms. To see that this is so, consider an experiment which is repeatedly performed under identical conditions, and let X_i be 1 or 0 according as a particular event A does or does not occur when the experiment is performed for the ith time. Then in the first N performances of the experiment the event A occurs

$$N(A) = \sum_{i=1}^{N} X_i$$

times. The random variables $X_1, X_2, \ldots$ are independent and identically distributed (this is the mathematical equivalent of the statement that the experiments are performed "under identical conditions"), and, if $P(A) = p$,

$$E(X_i) = 1 \times p + 0 \times (1-p) = p.$$

The weak law tells us that

$$\frac{N(A)}{N} \xrightarrow{P} p \quad \text{as} \quad N \to \infty,$$

and the strong law that

$$\frac{N(A)}{N} \to p \text{ a.s.} \quad \text{as} \quad N \to \infty.$$

Either assertion justifies our taking $N(A)/N$ for some large N as an estimate of p.

Thus with either law of large numbers, the wheel of probability has come full circle.

It is possible that the precise difference between the two laws may not yet be clear to the reader. For the random variables $X_1, X_2, \ldots$ just defined, the difference can perhaps best be seen by considering appropriate probability spaces.

For the weak law we consider, for each positive integer N, the sample space Ω consisting of the 2^N sequences of N terms, each term being either 0 or 1. The value of X_i at any point (i.e. sequence) of Ω is the ith term of the sequence. The appropriate sigma-algebra $\mathscr{F}$ consists of all subsets of Ω. The probability measure P is determined by its values for those sets which consist of a single point, such values being defined in the obvious manner, for example, the probability of $\{(1, 0, 1, 1, \ldots, 0, 1)\}$ is defined to be $pqpp \ldots qp$, where $q = 1-p$. In this sample space we consider the set consisting of those sequences for which

$$\left|\frac{N(A)}{N}-p\right| \geqslant \varepsilon$$

($N(A)$ is the number of 1's in the sequence), and compute its probability p_N. In fact, $p_N = \Sigma' \binom{N}{r} p^r q^{N-r}$, where Σ' denotes summation over those values of r for which $|(r/N)-p| \geqslant \varepsilon$. Finally, we verify that $p_N \to 0$ as $N \to \infty$. It is clear that, whatever computational difficulties may arise, no more is involved than an estimate of a sum of binomial probabilities.

For the strong law we consider the non-countable sample space Ω consisting of all infinite sequences of terms, each term being either 0 or 1. The X_i's are defined as before. We then construct (and herein lies one of the principal difficulties) or assume given (and this, in effect, is what we do in this book) a sigma-algebra $\mathscr{F}$ and a probability measure P having the appropriate properties. These "appropriate properties" must ensure that the set of all sequences with first term 1 is a set of $\mathscr{F}$ with probability p, and the set of all sequences with first term 1 and second term 0 is a set of $\mathscr{F}$ with probability pq, and so on. In this sample space we consider the set S consisting of all sequences for which

$$\frac{N(A)}{N} \to p \quad \text{as} \quad N \to \infty$$

($N(A)$ is the number of 1's in the first N terms of the sequence). We have to verify that $S \in \mathscr{F}$ and $P(S) = 1$.

The above considerations suggest that the strong law is the much deeper result, and this is confirmed by the fact that the weak law follows immediately from the strong law (by Theorem 1 (i)).

It may be of interest to state two results which are, in effect, the converses of the two laws of large numbers, and throw further light on the relation between them. We suppose that $X_1, X_2, \ldots$ are independent random variables which have the same distribution as some random variable X ($X_1, X_2, \ldots$ are *independent observations* of X), and, as before, let

$$\bar{X}_n = \frac{1}{n}(X_1+X_2+\ldots+X_n) \qquad (n = 1, 2, \ldots).$$

Firstly, it can be shown that there exists a real number c such that

$$\bar{X}_n \xrightarrow{P} c \quad \text{as} \quad n \to \infty \tag{9}$$

if and only if

$$nP(|X|) \geqslant n) \to 0 \quad \text{as} \quad n \to \infty \tag{10}$$

and

$$E(XI_{A_n}) \to c \quad \text{as} \quad n \to \infty, \tag{11}$$

where $A_n = \{|X| < n\}$ $(n = 1, 2, \ldots)$. The reader will easily verify that if $E(X)$ is finite, then (10) and (11) both hold with $c = E(X)$ (by Exercise 5.18 (ii) and the dominated convergence theorem respectively). However, if X is a random variable which takes the values $+k$ and $-k$ each with probability $b/(k^2 \log k)$ $(k = 2, 3, \ldots)$, where b is a suitable positive constant, it is easily seen that (10) and (11) both hold with $c = 0$, but that $E(X)$ does not exist. Thus the condition that (9) hold for some c is slightly weaker than the condition that $E(X)$ be finite.

Secondly, it can be shown that there exists a real number c such that

$$\bar{X}_n \to c \text{ a.s.} \quad \text{as} \quad n \to \infty \tag{12}$$

if and only if $E(X)$ is finite and equal to c. (The "if" part of the assertion is, of course, the strong law itself; for an outline of the proof of the "only if" part see Exercise 21.) Thus the condition that (12) hold for some c is that $E(X)$ be finite.

It follows that the strong law really is stronger than the weak law, since (12) holds for some c only if $E(X)$ is finite, whereas (9) holds under the weaker condition (10) and (11).

A digression on product spaces

In discussing the laws of large numbers, we have assumed that the independent and identically distributed random variables $X_1, X_2, \ldots$ are all defined on one and the same probability space $(\Omega, \mathscr{F}, P)$.

But suppose that we are given a random variable X defined on a probability space $(\Omega, \mathscr{F}, P)$. Can we construct a probability space $(\Omega^{(n)}, \mathscr{F}^{(n)}, P^{(n)})$, and define random variables $X_1, X_2, \ldots, X_n$ on it in such a way that $X_1, X_2, \ldots, X_n$ are independent and have the same distribution as X? The answer is "yes", and, without going into details we indicate how it can be done. Firstly, $\Omega^{(n)}$ is defined to be $\Omega \times \Omega \times \ldots \times \Omega$ (n factors), and so a typical point of $\Omega^{(n)}$ is $\omega^{(n)} = (\omega_1, \omega_2, \ldots, \omega_n)$, where $\omega_1, \omega_2, \ldots, \omega_n \in \Omega$. Secondly, if $\mathscr{C}$ is the class of all sets

$$A_1 \times A_2 \times \ldots \times A_n \quad (A_1, A_2, \ldots, A_n \in \mathscr{F}),$$

$\mathscr{F}^{(n)}$ is defined to be $\sigma(\mathscr{C})$. Thirdly, and this step is by no means trivial, it is possible to construct a probability measure $P^{(n)}$ on $\mathscr{F}^{(n)}$ with the property that

$$P^{(n)}(A_1 \times A_2 \times \ldots \times A_n) = P(A_1) F(A_2) \ldots P(A_n) \quad (A_1, A_2, \ldots, A_n \in \mathscr{F}).$$

Finally, let X_i be defined on $\Omega^{(n)}$ by

$$X_i(\omega^{(n)}) = X(\omega_i) \quad (\omega^{(n)} = (\omega_1, \omega_2, \ldots, \omega_n) \in \Omega^{(n)}).$$

Then it can easily be shown that $X_1, X_2, \ldots, X_n$ are independent, and have the same distribution as X.

A further question is whether, given the above random variable X, it is possible to construct a probability space $(\Omega^{(\infty)}, \mathscr{F}^{(\infty)}, P^{(\infty)})$ and define random variables $X_1, X_2, \ldots$ on it in such a way that $X_1, X_2, \ldots$ are independent and have the same distribution as X. Once again, the answer is "yes", and we outline the procedure. Firstly, $\Omega^{(\infty)}$ is defined to be $\Omega \times \Omega \times \ldots$ (a countably infinite product), and so a typical point of $\Omega^{(\infty)}$ is $\omega^{(\infty)} = (\omega_1, \omega_2, \ldots)$, where $\omega_1, \omega_2, \ldots \in \Omega$. Secondly, if $\mathscr{C}$ is the class of all sets

$$A_1 \times A_2 \times \ldots \ (A_1, A_2, \ldots \in \mathscr{F}),$$

$\mathscr{F}^{(\infty)}$ is defined to be $\sigma(\mathscr{C})$. Thirdly, and this step is even less trivial than before, it is possible to construct a probability measure $P^{(\infty)}$ on $\mathscr{F}^{(\infty)}$ with the property that

$$P^{(\infty)}(A_1 \times A_2 \times \ldots) = \lim_{n \to \infty} P(A_1) P(A_2) \ldots P(A_n) \quad (A_1, A_2, \ldots \in \mathscr{F})$$

(it is clear that the limit exists). Finally, let X_i be defined on $\Omega^{(\infty)}$ by

$$X_i(\omega^{(\infty)}) = X(\omega_i) \quad (\omega^{(\infty)} = (\omega_1, \omega_2, \ldots) \in \Omega^{(\infty)}).$$

Then it can easily be shown that $X_1, X_2, \ldots$ are independent, and have the same distribution as X.

Suppose that we are given a random variable X defined on a probability space $(\Omega, \mathscr{F}, P)$, and that $X_1, X_2, \ldots$ are independent observations of X. When discussing the weak law of large numbers we can, if we so desire, consider only the *finite-dimensional* probability spaces $\Omega^{(n)}$ $(n = 1, 2, \ldots)$, and regard the event $\{|\bar{X}_n - \mu| \geqslant \varepsilon\}$ as a set of $\mathscr{F}^{(n)}$. To discuss the strong law of large numbers, we have to consider the *infinite-dimensional* space $\Omega^{(\infty)}$, and regard the event $\{\bar{X}_n \to \mu \text{ as } n \to \infty\}$ as a set in $\mathscr{F}^{(\infty)}$. This shows, once again, that the strong law is a deeper result than the weak law.

We end this chapter with a final result on convergence in distribution which we shall need later. As a necessary preliminary, we prove the following lemma, since, although the result is widely known, the proof does not seem to be readily accessible in the literature.

Lemma 2. *Suppose that, for every positive integer k,*

$$a_{k1}, a_{k2}, \ldots$$

is a bounded sequence. Then there exist positive integers $n_1, n_2, \ldots$ such that $n_1 < n_2 < \ldots$ and every one of the subsequences

$$a_{kn_1}, a_{kn_2}, \ldots \qquad (k = 1, 2, \ldots)$$

is convergent.

Proof. The argument uses the "diagonal process" due to Cantor.

To avoid proliferating suffixes, let a_{mn} be written $a(m, n)$.

Since the sequence $a(1, n)$ $(n = 1, 2, \ldots)$ is bounded, it has a convergent subsequence, say

$$a(1, m(1, 1)), \quad a(1, m(1, 2)), \quad a(1, m(1, 3)), \ldots .$$

Since the sequence $a(2, m(1, n))$ $(n = 1, 2, \ldots)$ is bounded, it has a convergent subsequence, say

$$a(2, m(2, 1)), \quad a(2, m(2, 2)), \quad a(2, m(2, 3)), \ldots,$$

and so on.

Thus we have sequences of positive integers

$$S_r\colon \quad m(r, 1), \quad m(r, 2), \quad m(r, 3), \ldots \qquad (r = 1, 2, \ldots)$$

such that

(i) S_{r+1} is a subsequence of S_r, and
(ii) $a(r, m(r, n))$ tends to a limit b_r, say, as $n \to \infty$.

Let $n_r = m(r, r)$ $(r = 1, 2, \ldots)$. Then

$$\begin{aligned} n_{r+1} = m(r+1, r+1) &> m(r+1, r) \\ &\geqslant m(r, r) \quad \text{(by (i))} \\ &= n_r. \end{aligned}$$

Also

$$a_{kn_k}, a_{kn_{k+1}}, a_{kn_{k+2}}, \ldots,$$

i.e. $a(k, m(k, k)), \quad a(k, m(k+1, k+1)), \quad a(k, m(k+2, k+2)), \ldots,$

is, by (i), a subsequence of

$$a(k, m(k, k)), \quad a(k, m(k, k+1)), \quad a(k, m(k, k+2)), \ldots .$$

Since, by (ii) this last sequence is convergent with limit b_k, it follows that the sequence $a_{kn_1}, a_{kn_2}, a_{kn_3}, \ldots$ is convergent to the same limit.

Theorem 5 *Suppose that $F_1, F_2, \ldots$ are distribution functions satisfying the condition that to every $\varepsilon > 0$ there corresponds a number $t = t(\varepsilon) > 0$ such that*

$$F_n(t) - F_n(-t) > 1 - \varepsilon \quad \textit{for all sufficiently large } n.$$

Then there exists a subsequence $F_{n_1}, F_{n_2}, \ldots$ of $F_1, F_2, \ldots$ and a distribution function F such that

$$F_{n_p} \xrightarrow{D} F \quad \text{as} \quad p \to \infty.$$

Note. Some such condition as that given is necessary to prevent the distribution "disappearing to infinity", in which case no limiting distribution function exists. If, say, $F_n(x) = 0$ if $x < n$, and $= 1$ if $x \geqslant n$, then $F_n(x) \to 0$ as $n \to \infty$ $(x \in R)$, and the conclusion of the theorem cannot hold.

Proof. Let the rational numbers, written in some order as a sequence, be $r_1, r_2, \ldots$. For every positive integer k, the sequence

$$F_1(r_k), \; F_2(r_k), \ldots$$

is bounded, and so, by Lemma 2, there exist positive integers $n_1, n_2, \ldots$ such that $n_1 < n_2 < \ldots$ and every one of the subsequences

$$F_{n_1}(r_k), \; F_{n_2}(r_k), \ldots \qquad (k = 1, 2, \ldots)$$

is convergent.

For each rational number r let

$$G(r) = \lim_{p\to\infty} F_{n_p}(r).$$

Then

$$0 \leqslant G(r) \leqslant 1 \quad \text{for every rational number } r. \tag{13}$$

We now define $F(x)$ for any real number x to be the greatest lower bound of those numbers $G(r)$ for which r is rational and $> x$. To show that F is the desired distribution function, we have to establish the following results:

(a) F is non-decreasing on R;
(b) F is everywhere continuous on the right;
(c) $F(x) \to 0$ as $x \to -\infty$ and $F(x) \to 1$ as $x \to \infty$;
(d) $F_{n_p}(x) \to F(x)$ as $p \to \infty$ whenever x is a point of continuity of F.

It follows immediately from (13) that

$$0 \leqslant F(x) \leqslant 1 \quad \text{for every real number } x,$$

and from the definition of F that

$$F(x) \leqslant F(y) \quad \text{for all real numbers } x \text{ and } y \text{ with } x \leqslant y$$

(and so (a) holds).

To prove (b), let x be any real number, and suppose that $\varepsilon > 0$. By the definition of F, there exists a rational number $r > x$ such that

$$G(r) < F(x)+\varepsilon.$$

But

$$F(y) \leqslant G(r) \quad (y < r)$$

(again by the definition of F), and so

$$F(x) \leqslant F(y) < F(x)+\varepsilon \quad (x \leqslant y < r).$$

Therefore F is continuous on the right at x.

To prove (c) we shall need to use the condition of the theorem.

Suppose that $\varepsilon > 0$. Then, by that condition, there exists a number $t > 0$ such that

$$F_{n_p}(t)-F_{n_p}(-t) > 1-\varepsilon \quad \text{for all sufficiently large } p.$$

For such p, and any rational number $r > t$,

$$F_{n_p}(r) \geqslant F_{n_p}(t) > 1-\varepsilon+F_{n_p}(-t) \geqslant 1-\varepsilon,$$

and so, by the definition of G,

$$G(r) \geqslant 1-\varepsilon.$$

It follows that

$$1 \geqslant F(x) \geqslant 1-\varepsilon \quad (x > t)$$

(by the definition of F), and so $F(x) \to 1$ as $x \to \infty$.

A similar argument shows that

$$0 \leqslant F(x) \leqslant \varepsilon \quad (x < -t),$$

and so $F(x) \to 0$ as $x \to -\infty$.

We show finally that (d) holds.

Suppose that F is continuous at x. If r is rational and $> x$, $F_{n_p}(x) \leqslant F_{n_p}(r)$ (all p), and so

$$\overline{\lim_{p\to\infty}}\, F_{n_p}(x) \leqslant \lim_{p\to\infty} F_{n_p}(r) = G(r).$$

This holds for every rational number $r > x$, and so, by the definition of $F(x)$,

$$\overline{\lim_{p\to\infty}}\, F_{n_p}(x) \leqslant F(x). \tag{14}$$

Now let y be any real number $< x$. Choose a rational number r such that $y < r < x$. Since $F_{n_p}(r) \leqslant F_{n_p}(x)$ (all p), it follows, on letting p tend to infinity, that

$$G(r) \leqslant \underline{\lim_{p\to\infty}}\, F_{n_p}(x)$$

and so, by the definition of $F(y)$,

$$F(y) \leqslant \underline{\lim_{p\to\infty}}\, F_{n_p}(x).$$

This holds for every real number $y < x$, and so, since F is assumed to be continuous at x,

$$F(x) \leqslant \underline{\lim_{p\to\infty}}\, F_{n_p}(x). \tag{15}$$

Therefore, by (14) and (15),

$$F_{n_p}(x) \to F(x) \quad \text{as} \quad p \to \infty,$$

and (d) is established.

Exercises

1. (i) Suppose that $X, Y, Z \in L_r$ (see page 112). Prove that

$$E(|X-Y|^r) \leqslant c_r\{E(|X-Z|^r)+E(|Y-Z|^r)\},$$

where $c_r = \max(1, 2^{r-1})$.

(ii) Let X, Y, Z be any random variables. Prove that

$$P(|X-Y| \geqslant \varepsilon) \leqslant P(|X-Z| \geqslant \tfrac{1}{2}\varepsilon)+P(|Y-Z| \geqslant \tfrac{1}{2}\varepsilon) \quad (\varepsilon > 0).$$

Hint. $$\{|X-Y| \geqslant \varepsilon\} \subseteq \{|X-Z| \geqslant \tfrac{1}{2}\varepsilon\}+\{|Y-Z| \geqslant \tfrac{1}{2}\varepsilon\}.$$

Note. These two useful inequalities will often be needed for the solutions of the following exercises.

2. (i) Suppose that $X_n \xrightarrow{r} X$ as $n \to \infty$, where $r > 0$. Prove that

$$E(|X_m-X_n|^r) \to 0 \quad \text{as} \quad m, n \to \infty.$$

(ii) Suppose that $X_n \xrightarrow{P} X$ as n $\to \infty$. Prove that

$$P(|X_m-X_n| \geqslant \varepsilon) \to 0 \quad \text{as} \quad m, n \to \infty \quad (\varepsilon > 0).$$

3. Suppose that $X_n \xrightarrow{s} X$ as $n \to \infty$, and that $0 < r < s$. Prove that $X_n \xrightarrow{r} X$ as $n \to \infty$.

Hint. Use the inequality of Exercise 5.4.

Hence show that $E(X_n) \to E(X)$ as $n \to \infty$ if $s \geqslant 1$.

4. (a) Suppose that X_1, X_2, ... are non-negative random variables for which $\sum_{n=1}^{\infty} E(X_n)$ is convergent. Prove that $\sum_{n=1}^{\infty} X_n$ is a.s. convergent.

Hint. Let $B = \left\{\omega\colon \sum_{n=1}^{\infty} X_n(\omega) \text{ is divergent}\right\}$. Proceed as in the solution of Exercise 5.21 to show that

$$kP(B) \leqslant \sum_{n=1}^{\infty} E(X_n)$$

for every positive integer k.

(b) Suppose that $X_1, X_2, \ldots, X \in L_r$, and that

$$\sum_{n=1}^{\infty} E(|X_n-X|^r)$$

is convergent. Prove that $X_n \to X$ a.s. as $n \to \infty$.

Note. Clearly $X_n \to X$ as $n \to \infty$.

5. (a) Suppose that X_1, X_2, ..., X are random variables for which

$$\sum_{n=1}^{\infty} P(|X_n-X| \geqslant \varepsilon) \text{ is convergent} \quad (\varepsilon > 0).$$

Prove that $X_n \to X$ a.s. as $n \to \infty$.

Note. Clearly $X_n \xrightarrow{P} X$ as $n \to \infty$.

Hint. A particular case of this result was used in Exercise 6.15. Steps (iv)–(vi) of the solution outlined there apply, with obvious modifications, to the present exercise.

(b) Suppose that the random variables $X_1, X_2, \ldots$ are independent, and that $X_n \to 0$ a.s. as $n \to \infty$. Prove that $\sum_{n=1}^{\infty} P(|X_n| \geqslant \varepsilon)$ is convergent ($\varepsilon > 0$).

Hint. Use the appropriate Borel–Cantelli lemma.

6. Suppose that $X_1, X_2, \ldots$ are independent and identically distributed random variables with finite second moment. Prove that

$$\frac{1}{n}(X_1+X_2+\ldots+X_n) \xrightarrow{2} E(X_1) \quad \text{as} \quad n \to \infty.$$

Compare this result with that of Exercise 6.14.

7. (a) Suppose that $X_n \xrightarrow{P} X$ and $Y_n \xrightarrow{P} Y$ as $n \to \infty$. Prove that

$$X_n+Y_n \xrightarrow{P} X+Y \quad \text{as} \quad n \to \infty.$$

(b) State and prove the corresponding result for convergence in rth mean.

8. (a) Suppose that $X_n \xrightarrow{P} 0$ as $n \to \infty$. Prove that, for any random variable Y, $X_nY \xrightarrow{P} 0$ as $n \to \infty$.

Hint. Show that, for any $k > 0$,

$$P(|X_nY| \geqslant \varepsilon) \leqslant P(|X_n| \geqslant \varepsilon/k)+P(|Y| \geqslant k),$$

and hence that

$$\overline{\lim_{n\to\infty}} P(|X_nY| \geqslant \varepsilon) \leqslant P(|Y| \geqslant k).$$

Now let k tend to infinity.

(b) Suppose that $X_n \xrightarrow{P} 0$ and $Y_n \xrightarrow{P} 0$ as $n \to \infty$. Prove that $X_nY_n \xrightarrow{P} 0$ as $n \to \infty$.

(c) Suppose that $X_n \xrightarrow{P} X$ and $Y_n \xrightarrow{P} Y$ as $n \to \infty$. Prove that $X_nY_n \xrightarrow{P} XY$ as $n \to \infty$.

Hint. $X_nY_n - XY = (X_n-X)(Y_n-Y)+Y(X_n-X)+X(Y_n-Y)$.

Note. Thus, in particular, if $X_n \xrightarrow{P} X$ and $c_n \to c$ as $n \to \infty$, where $c_1, c_2, \ldots, c \in R$, then

$$c_nX_n \xrightarrow{P} cX \quad \text{as} \quad n \to \infty.$$

9. Suppose that $X_n \xrightarrow{r} X$ and $c_n \to c$ as $n \to \infty$, where $c_1, c_2, \ldots, c \in R$. Prove that

$$c_n X_n \xrightarrow{r} cX \quad \text{as} \quad n \to \infty.$$

10. Suppose that $X_n \xrightarrow{P} X$ as $n \to \infty$, and that $f: R \to R$ is continuous on R. Prove that $f(X_n) \xrightarrow{P} f(X)$ as $n \to \infty$.

Hint. Suppose that $\varepsilon > 0$ and $\delta > 0$. Choose real numbers a, b with $a < b$ such that

(i) $F(a) < \delta$ and $F(b) > 1-\delta$;
(ii) F is continuous at a and b (see Lemma 1, corollary).

Hence, by Theorem 2, $F_n(a) < \delta$ and $F_n(b) > 1-\delta$ for all sufficiently large n. For such n

$$P\{|f(X_n)-f(X)| \geqslant \varepsilon\}$$
$$< P\{|f(X_n)-f(X)| \geqslant \varepsilon \quad \text{and} \quad X_n, X \in [a, b]\}+4\delta.$$

The result follows since f is uniformly continuous on $[a, b]$.

Note. It can be shown similarly that if $X_{kn} \xrightarrow{P} X_k$ as $n \to \infty$ $(k = 1, 2, \ldots, m)$, and f: $R^m \to R$ is continuous on R^m, then

$$f(X_{1n}, X_{2n}, \ldots, X_{mn}) \xrightarrow{P} f(X_1, X_2, \ldots, X_m) \quad \text{as} \quad n \to \infty.$$

Exercises 7(a) and 8(c) are particular cases of this result.

11. Suppose that $X_1, X_2, \ldots, X$ are non-negative random variables for which $X_n \xrightarrow{P} X$ as $n \to \infty$. Prove that $\sqrt{X_n} \xrightarrow{P} \sqrt{X}$ as $n \to \infty$.

Hint. Use Exercise 10. Alternatively, give a direct proof using the inequality

$$|\sqrt{a}-\sqrt{b}| \leqslant \sqrt{|a-b|} \quad (a, b \geqslant 0).$$

12. (a) For any two random variables X and Y let

$$d(X, Y) = E\left(\frac{|X-Y|}{1+|X-Y|}\right).$$

Show that, for any random variables X, Y, Z,

(i) $0 \leqslant d(X, Y) = d(Y, X) < 1$;
(ii) $d(X, Y) = 0$ if and only if $X = Y$ a.s.;
(iii) $d(X, Z) \leqslant d(X, Y)+d(Y, Z)$.

Note. Thus if two random variables which are equal a.s. are regarded as identical, d is a metric on the space of all random variables defined on a given probability space $(\Omega, \mathscr{F}, P)$.

(b) Show that $X_n \xrightarrow{P} X$ as $n \to \infty$ if and only if $d(X_n, X) \to 0$ as $n \to \infty$.

Hint. Use Exercise 5.10 with $f(x) = \dfrac{x}{1+x}$ $(x \geqslant 0)$ to show that

$$\frac{\varepsilon}{1+\varepsilon} P(|X_n - X|) \geqslant \varepsilon) \leqslant d(X_n, X) \leqslant \frac{\varepsilon}{1+\varepsilon} + P(|X_n - X| \geqslant \varepsilon).$$

13. Suppose that the random variable X_n is a constant c_n on Ω ($n = 0, 1, 2, \ldots$). Prove that $X_n \xrightarrow{D} X_0$ as $n \to \infty$ if and only if $c_n \to c_0$ as $n \to \infty$.

14. Suppose that $F_1, F_2, \ldots, F$ and G are distribution functions, and that $F_n \xrightarrow{D} F$ and $F_n \xrightarrow{D} G$ as $n \to \infty$. Prove that $F = G$.

15. Suppose that the distribution function $F_n(x) = 0$ if $x \leqslant 0$, $= x^n$ if $0 \leqslant x \leqslant 1$, and $= 1$ if $x \geqslant 1$ ($n = 1, 2, \ldots$). Find the distribution function F for which $F_n \xrightarrow{D} F$ as $n \to \infty$.

16. Suppose that $X_n \xrightarrow{D} X$ as $n \to \infty$, and that the distribution function F of X is continuous at a and b, where $a < b$. Prove that

$$P(a < X_n < b) \to F(b) - F(a) \quad \text{as} \quad n \to \infty$$

and

$$P(a \leqslant X_n \leqslant b) \to F(b) - F(a) \quad \text{as} \quad n \to \infty.$$

17. Prove the "if" part of Theorem 4.

Hint. Suppose that $\varepsilon > 0$ and that $c \in R$. Let $f(x) = 1$ if $x \leqslant c$, $= (c+\varepsilon-x)/\varepsilon$ if $c \leqslant x \leqslant c+\varepsilon$, and $= 0$ if $x \geqslant c+\varepsilon$. Show that

$$F_n(c) \leqslant E\{f(X_n)\} \quad (\text{all } n) \quad \text{and} \quad F(c+\varepsilon) \geqslant E\{f(X)\},$$

and hence that

$$\overline{\lim_{n \to \infty}} F_n(c) \leqslant F(c+\varepsilon).$$

Show similarly that $\underline{\lim}_{n \to \infty} F_n(c) \geqslant F(c-\varepsilon)$, and deduce that $F_n(c) \to F(c)$ as $n \to \infty$ if F is continuous at c.

18. Suppose that $X_n \xrightarrow{D} X$ as $n \to \infty$, and that $c \in R$. Prove that

(i) $X_n + c \xrightarrow{D} X + c$ as $n \to \infty$;
(ii) $cX_n \xrightarrow{D} cX$ as $n \to \infty$.

Hint. It suffices to prove (ii) for $c > 0$ and for $c = -1$. A direct proof that

$$-X_n \xrightarrow{D} -X \quad \text{as} \quad n \to \infty$$

is straightforward, but the details are a bit messy. It is simpler to use Theorem 4.

19. (a) Suppose that $X_n \xrightarrow{D} X$ and $Y_n \xrightarrow{P} 0$ as $n \to \infty$.

(i) Prove that $X_n + Y_n \xrightarrow{D} X$ as $n \to \infty$.

Hint. See the proof of Theorem 2.

(ii) Prove that $X_n Y_n \xrightarrow{P} 0$ as $n \to \infty$.

Hint. Suppose that $\varepsilon > 0$. Choose $k > 0$ such that F is continuous at both k and $-k$. Show that

$$P(|X_n Y_n| \geqslant \varepsilon) \leqslant P(|Y_n| \geqslant \varepsilon/k) + F_n(-k) + 1 - F_n(k).$$

Hence $\overline{\lim_{n \to \infty}} P(|X_n Y_n| \geqslant \varepsilon) \leqslant F(-k) + 1 - F(k)$.

Now let k tend to infinity.

(b) Suppose that $X_n \xrightarrow{D} X$ and $Y_n \xrightarrow{P} c$ as $n \to \infty$, where $c \in R$. Prove that $X_n + Y_n \xrightarrow{D} X + c$ and $X_n Y_n \xrightarrow{D} cX$ as $n \to \infty$.

Hint. Use (a) and Exercise 18.

20. Suppose that X is a random variable for which $E(X^2)$ is finite. Let $X_1, X_2, \ldots$ be independent observations of X, and let

$$\overline{X}_n = \frac{1}{n}(X_1 + X_2 + \ldots + X_n) \qquad (n = 1, 2, \ldots).$$

Use the strong law of large numbers to show that

$$\frac{1}{n} \sum_{i=1}^{n} (X_i - \overline{X}_n)^2 \to \operatorname{var} X \text{ a.s.} \quad \text{as} \quad n \to \infty.$$

21. Let $X_1, X_2, \ldots$ be independent observations of a random variable X, and suppose that there exists a real number c such that

$$\frac{1}{n} \sum_{k=1}^{n} X_k \to c \text{ a.s.} \quad \text{as} \quad n \to \infty.$$

Prove the following:

(i) $\frac{1}{n} X_n \to 0$ a.s. as $n \to \infty$.

(ii) $\sum_{n=1}^{\infty} P(|X_n| \geqslant n)$ is convergent.

Hint. Use Exercise 5(b).

(iii) $E(X)$ is finite.

Hint. Use Exercise 5.18(i).

(iv) $E(X) = c$.

The following exercise illustrates the principle that random variables which are almost surely equal may be regarded as identical for the purposes of probability theory (see also Exercise 5.5).

22. Suppose that $X_1, X_2, \ldots, X, Y_1, Y_2, \ldots, Y$ are random variables.

(a) (i) Show that if $X_n \to X$ a.s. as $n \to \infty$

and $X_n \to Y$ a.s. as $n \to \infty$

then $X = Y$ a.s.

(ii) Show that if $X_n \to X$ a.s. as $n \to \infty$,

$Y_n = X_n$ a.s. $(n = 1, 2, \ldots)$,

and $Y = X$ a.s.,

then $Y_n \to Y$ a.s. as $n \to \infty$.

(b) Show that the results of (a) still hold if "$\to$ ... a.s." is replaced throughout by "$\xrightarrow{r}$" or by "$\xrightarrow{P}$".

(c) Show that if "$\to$ ... a.s." is replaced throughout by "$\xrightarrow{D}$" then only the analogue of (a) (ii) still holds. (See Exercise 14 for the analogue of (a) (i).)

23. Suppose that $X_n \to X$ a.s. as $n \to \infty$, and that $X_n \xrightarrow{r} Y$ as $n \to \infty$. Prove that $X = Y$ a.s.

Hint. Use Theorem 1 and Exercise 22(b) ("$\xrightarrow{P}$").

8 Characteristic functions

It has already been established that the distribution function of a random variable X determines the distribution of X, and so any probability theory question relating to X can, in principle, be answered when its distribution function is known. More generally, if $X_1, X_2, \ldots, X_n$ are random variables and $\mathbf{X} = (X_1, X_2, \ldots, X_n)$, the distribution function of $\mathbf{X}$ (see Exercise 4.12) determines the distribution of $\mathbf{X}$, i.e. the function $P_{\mathbf{X}}: \mathscr{B}^n \to [0, 1]$ defined by

$$P_{\mathbf{X}}(B) = P\{\mathbf{X}^{-1}(B)\} \quad (B \in \mathscr{B}^n)$$

(see Exercise 4.1 (ii)), and so, in principle, the answer to any probability theory question relating to $\mathbf{X}$, for example, whether $E(X_1 X_2)$ is or is not finite. In particular, the distribution function of $\max(X_1, X_2, \ldots, X_n)$ is easily obtained from that of $\mathbf{X}$.

Many frequently occurring and important questions relate to the sum Y of independent random variables $X_1, X_2, \ldots, X_n$. If X_i has the distribution function F_i $(i = 1, 2, \ldots, n)$, then the distribution function $F: R^n \to R$ of $\mathbf{X} = (X_1, X_2, \ldots, X_n)$ is given by

$$F(x_1, x_2, \ldots, x_n) = F_1(x_1) F_2(x_2) \ldots F_n(x_n) \quad (x_1, x_2, \ldots, x_n \in R).$$

Unfortunately, the distribution function G of Y is not easily obtainable from F. For example, when $n = 2$, it can be shown that

$$G(x) = E\{F_2(x - X_1)\} \quad (x \in R),$$

and the formulae for larger values of n are correspondingly more complex.

Thus in order to deal with problems relating to sums of random variables a new analytical tool is required, and this is provided by the theory of characteristic functions. The characteristic function of a random variable X is the function φ defined on R by

$$\varphi(t) = E(e^{itX}) = E(\cos tX) + iE(\sin tX)$$

(see Definition 5.4). Thus φ is a complex-valued function of a real variable. It will be shown that the characteristic function φ of X has many desirable properties; for example, it is always everywhere continuous (unlike the distribution function of X), and it determines (or characterises) the distribution function, and therefore the distribution, of X. It will also be shown that if $X_1, X_2, \ldots, X_n$ are independent random variables with characteristic functions $\varphi_1, \varphi_2, \ldots, \varphi_n$ respectively, then the characteristic function φ of $X_1+X_2+ + \ldots +X_n$ is given very simply by

$$\varphi(t) = \varphi_1(t)\,\varphi_2(t)\,\ldots\,\varphi_n(t) \quad (t \in R).$$

Thus characteristic functions provide a powerful tool for dealing with sums of independent random variables. Furthermore, it will be shown that if $X_1, X_2, \ldots, X$ are random variables with corresponding characteristic functions $\varphi_1, \varphi_2, \ldots, \varphi$, then

$$X_n \xrightarrow{D} X \quad \text{as} \quad n \to \infty$$

if and only if

$$\varphi_n(t) \to \varphi(t) \quad \text{as} \quad n \to \infty \quad (t \in R).$$

In what follows, we shall often be dealing with expected values of "complex-valued" random variables. We remind the reader of the convention we adopted in Chapter 5, namely that the "complex-valued" form of a theorem will be denoted by the affix C.

Let X be a random variable. Then for any real number t, $\cos tX$ (i.e. $\cos(tX)$, not $(\cos t)X$) and $\sin tX$ are bounded random variables (by Theorems 3.2 (i) and 3.5), and so $E(\cos tX)$ and $E(\sin tX)$ are finite (by Theorem 5.4, corollary). Therefore

$$E(e^{itX}) = E(\cos tX)+iE(\sin tX)$$

is finite (see Definition 5.4).

Definition. The *characteristic function* φ of a random variable X is the function defined on R by

$$\varphi(t) = E(e^{itX}) \quad (t \in R).$$

Certain results relating to the function e^{it} $(t \in R)$ will be of use later. They are contained in the following lemma.

Lemma 1

(i) $$\left| e^{it} - \sum_{k=0}^{n-1} \frac{1}{k!}(it)^k \right| \leqslant \frac{1}{n!}|t|^n \quad (t \in R;\ n = 1, 2, \ldots).$$

(ii) *For real t and* $n = 1, 2, \ldots$ *let* $g_n(t)$ *be defined by*

$$e^{it} = \sum_{k=0}^{n-1} \frac{1}{k!}(it)^k + \frac{1}{n!}(it)^n\{1+g_n(t)\}$$

if $t \neq 0$, *and let* $g_n(0) = 0$. *Then* g_n *is continuous and* $|g_n| \leqslant 2$ *on R.*

Proof (i). The inequality holds for $n = 1$, since

$$|e^{it}-1| = \left|\int_0^t ie^{iu}\,du\right| \leqslant \left|\int_0^t |ie^{iu}|\,du\right| = |t|.$$

A similar argument serves to complete a proof by induction over n.
(ii) Note that, by (i),

$$\left|\frac{(it)^n}{n!} g_n(t)\right| \leqslant \frac{|t|^{n+1}}{(n+1)!} \quad (t \in R),$$

which shows that g_n is continuous at 0 (clearly g_n is continuous on $R-\{0\}$), and

$$\left|\frac{(it)^n}{n!}\{1+g_n(t)\}\right| \leqslant \frac{|t|^n}{n!} \quad (t \in R),$$

which shows that $|g_n| \leqslant 2$ on $R-\{0\}$.

Theorem 1 *Suppose that* φ *is the characteristic function of a random variable X. Then the following results hold:*

(i) $\varphi(0) = 1$.
(ii) $|\varphi(t)| \leqslant 1 \quad (t \in R)$.
(iii) φ *is continuous on R.*
(iv) *If* $E(X^n)$ *is finite, where n is a positive integer, then, for* $k = 1, 2, \ldots, n$, $\varphi^{(k)}$ *is finite and continuous on R and*

$$\varphi^{(k)}(t) = i^k E(X^k e^{itX}) \quad (t \in R).$$

(v) *If* $E(X^n)$ *is finite, where n is a positive integer, then*

$$\varphi(t) = \sum_{k=0}^{n} \frac{1}{k!}\mu'_k i^k t^k + h(t) \quad (t \in R),$$

where $\mu'_k = E(X^k) \quad (k = 0, 1, \ldots, n)\,(\mu'_0 = 1)$ *and*

$$t^{-n}h(t) \to 0 \quad \text{as} \quad t \to 0.$$

Proof

(i) This is trivial.

(ii) $|\varphi(t)| = |E(e^{itX})|$

$$\leqslant E(|e^{itX}|) \quad \text{(by Theorem 5.4(ii)C)}$$

$$= 1.$$

(iii) This follows from Theorem 5.7C on taking $I = R$, $Y = 1$, and $X(t;\omega) = e^{itX(\omega)}$ (or, equivalently, $X^{(t)} = e^{itX}$).

(iv) We proceed by induction over k.

Since $E(X^n)$ is finite, it follows that $E(|X|^k)$ is finite for $k = 1, 2, \ldots, n$ (by Exercise 5.4). Therefore, in particular, $E(|X|)$ is finite. It now follows from Theorem 5.8C (with $I = R$, $Y = |X|$, and $X^{(t)} = e^{itX}$) that φ' is finite on R and

$$\varphi'(t) = E(iXe^{itX}) \quad (t \in R).$$

Theorem 5.7C now shows that φ' is continuous on R. Thus the result holds for $k = 1$.

Now suppose that the result holds for $k = m\,(< n)$. Then, again from Theorem 5.8C (with $I = R$, $Y = |X|^{m+1}$, and $X^{(t)} = i^m X^m e^{itX}$), it follows that $\varphi^{(m+1)}$ is finite on R and

$$\varphi^{(m+1)}(t) = E(i^{m+1}X^{m+1}e^{itX}) \quad (t \in R).$$

Once again an application of Theorem 5.7C shows that $\varphi^{(m+1)}$ is continuous on R. Thus the result holds for $k = m+1$ if it holds for $k = m$, and the proof is complete.

A direct proof of the continuity of $\varphi^{(k)}$ is, in fact, needed only for $k = n$. Its continuity for values of $k < n$ is an immediate consequence of its differentiability.

(v) *First proof*. With the notation of Lemma 1(ii)

$$h(t) = E\left\{\frac{1}{n!}(itX)^n g_n(tX)\right\}.$$

Therefore

$$|h(t)| \leqslant \frac{|t|^n}{n!} E\{|X|^n |g_n(tX)|\}$$

(by Theorem 5.4(ii)C). By Theorem 5.7C (with $I = R$, $Y = 2|X|^n$, and $X^{(t)} = |X|^n |g_n(tX)|$), $E\{|X|^n |g_n(tX)|\}$ is continuous on R, and so tends to its value at 0, which is 0, as $t \to 0$.

Second proof. By (iv)

$$\varphi^{(k)}(0) = i^k \mu_k' \qquad (k = 1, 2, \ldots, n).$$

Now if f is a function defined on some interval of the real line with centre at the origin, and $f^{(n)}(0)$ is finite, then, as $t \to 0$,

$$f(t) = f(0)+tf'(0)+\frac{t^2}{2!}f''(0)+ \dots +\frac{t^{n-1}}{(n-1)!}f^{(n-1)}(0)+\frac{t^n}{n!}\{f^{(n)}(0)+o(1)\}.$$

(see G. H. Hardy, *A Course of Pure Mathematics* (10th edn.), pp. 289–90 (Cambridge University Press, 1963) for a proof for real-valued f, and extend the result to complex-valued f by considering real and imaginary parts separately). The result follows immediately.

Notes. (a) It can be shown that φ is uniformly continuous on R (see Exercise 6).

(b) It is possible to establish a partial converse to part (iv) of the theorem, namely, that if $\varphi^{(n)}(0)$ is finite, where n is *even*, then $E(X^n)$ is finite (see Exercise 7 for the case $n = 2$). This does not necessarily hold if n is odd. For example, let X be the random variable which takes the values $+k$ and $-k$ each with probability $b/(k^2 \log k)$ $(k = 2, 3, \dots)$, where b is a suitable positive constant. Then its characteristic function φ is given by

$$\varphi(t) = 2b \sum_{k=2}^{\infty} (\cos kt)/(k^2 \log k) \quad (t \in R),$$

and so φ has a continuous derivative on R. However, $E(X)$ does not exist.

Theorem 2 (i) *Suppose that X and Y are independent random variables with characteristic functions φ and ψ respectively. Then $X+Y$ has the characteristic function $\varphi\psi$.*

(ii) *More generally, suppose that $X_1, X_2, \dots, X_n$ are independent random variables with characteristic functions $\varphi_1, \varphi_2, \dots, \varphi_n$ respectively. Then $X_1+X_2+ \dots +X_n$ has the characteristic function $\varphi_1\varphi_2 \dots \varphi_n$.*

Proof (i) Let χ denote the characteristic function of $X+Y$. Then, for all $t \in R$,

$$\begin{aligned}\chi(t) &= E\{e^{it(X+Y)}\}\\ &= E\{e^{itX}e^{itY}\}\\ &= E\{(\cos tX+i\sin tX)(\cos tY+i\sin tY)\}\\ &= E(\cos tX\cos tY)-E(\sin tX\sin tY)+iE(\cos tX\sin tY)\\ &\quad +iE(\sin tX\cos tY). \qquad (1)\end{aligned}$$

Now $\cos tX$ and $\cos tY$ are independent (by Theorem 6.1 (a)), and so

$$E(\cos tX\cos tY) = E(\cos tX)E(\cos tY)$$

(by Theorem 6.3). A similar argument can be applied to the other terms on the right-hand side of (1), and so

$$\begin{aligned}\chi(t) &= \{E(\cos tX)+iE(\sin tX)\}\{E(\cos tY)+iE(\sin tY)\} \\ &= \varphi(t)\,\psi(t),\end{aligned}$$

and the proof is complete.

(ii) This part is proved similarly, using Exercise 6.7 (i) and Theorem 6.3, corollary.

The next theorem and its corollary show that the characteristic function of a random variable determines its distribution function.

Theorem 3 (*the inversion formula*) *Suppose that X is a random variable with distribution function F and characteristic function φ, and let*

$$\tilde{F}(x) = \tfrac{1}{2}\{F(x-)+F(x)\} \quad (x \in R).$$

Then, for any real numbers a and b,

$$\tilde{F}(b)-\tilde{F}(a) = \lim_{c\to\infty} I(c),$$

where

$$I(c) = \frac{1}{2\pi}\int_{-c}^{c} \frac{e^{-iat}-e^{-ibt}}{it}\varphi(t)\,dt \quad (c>0). \tag{2}$$

Notes. (a) $\tilde{F}(a) = F(a)$ if and only if F is continuous at a.

(b) $(e^{-iat}-e^{-ibt})/(it)$ is defined to be $b-a$ when $t = 0$, and is therefore continuous on R. The integral in (2) is the Riemann integral of an everywhere continuous function.

(c) The function $(e^{-iat}-e^{-ibt})/(it)$ is bounded on R. In fact, for all $t \in R$,

$$\begin{aligned}\left|\frac{e^{-iat}-e^{-ibt}}{it}\right| &= \left|\frac{e^{\frac{1}{2}i(b-a)t}-e^{\frac{1}{2}i(a-b)t}}{it}\right| \\ &= \left|\frac{2\sin\frac{1}{2}(b-a)t}{t}\right| \leqslant |b-a|,\end{aligned}$$

since $|\sin x| \leqslant |x|$ for every real number x. The same inequality also follows from Lemma 1(*i*) with $n = 1$.

(d) What are, in effect, useful special cases of the inversion formula are given in Exercises 14 and 15.

Proof. There is no loss of generality in assuming that $a < b$.

By the definition of φ

$$I(c) = \frac{1}{2\pi}\int_{-c}^{c} \frac{e^{-iat}-e^{-ibt}}{it} E(e^{itX})\,dt$$

$$= \frac{1}{2\pi}\int_{-c}^{c} E\left\{\frac{e^{it(X-a)}-e^{it(X-b)}}{it}\right\} dt.$$

Therefore, by Theorem 5.9C, with $I = [-c, c]$, $Y = b-a$, and

$$X^{(t)} = \frac{1}{it}\{e^{it(X-a)}-e^{it(X-b)}\},$$

$$I(c) = E\left\{\frac{1}{2\pi}\int_{-c}^{c} \frac{e^{it(X-a)}-e^{it(X-b)}}{it}\,dt\right\}.$$

Now

$$\frac{1}{2\pi}\int_{-c}^{c} \frac{e^{it(X-a)}-e^{it(X-b)}}{it}\,dt$$

$$= \frac{1}{2\pi i}\int_{-c}^{c} \frac{\cos t(X-a)-\cos t(X-b)}{t}\,dt$$

$$+\frac{1}{2\pi}\int_{-c}^{c} \frac{\sin t(X-a)-\sin t(X-b)}{t}\,dt.$$

The first term vanishes, since the integrand is an odd function of t which is everywhere continuous and vanishes at the origin. The second term

$$= \frac{1}{\pi}\int_{0}^{c} \frac{\sin t(X-a)-\sin t(X-b)}{t}\,dt \quad \text{(because the integrand is an even function of } t\text{)}$$

$$= \frac{1}{\pi}\int_{0}^{c(X-a)} \frac{\sin u}{u}\,du - \frac{1}{\pi}\int_{0}^{c(X-b)} \frac{\sin u}{u}\,du$$

$$= \psi\{c(X-a)\}-\psi\{c(X-b)\},$$

where

$$\psi(t) = \frac{1}{\pi}\int_{0}^{t} \frac{\sin u}{u}\,du \quad (t \in R).$$

Thus

$$I(c) = E[\psi\{c(X-a)\}-\psi\{c(X-b)\}].$$

Now ψ is continuous on R, and

$$\psi(t) \to \begin{cases} \tfrac{1}{2} & \text{as} \quad t \to \infty \\ -\tfrac{1}{2} & \text{as} \quad t \to -\infty. \end{cases} \tag{3}$$

Therefore ψ is bounded on R, and so there exists a number k such that

$$|\psi(t)| \leqslant k \quad (t \in R). \tag{4}$$

Furthermore, it follows from (3) that, as $c \to \infty$,

$$\psi\{c(X-a)\}-\psi\{c(X-b)\} \to \begin{cases} 0 & \text{if} \quad X < a \quad \text{or} \quad X > b \\ \frac{1}{2} & \text{if} \quad X = a \quad \text{or} \quad X = b \\ 1 & \text{if} \quad a < X < b. \end{cases}$$

Let Z be the random variable defined to be 0 if $X < a$ or $X > b$, to be $\frac{1}{2}$ if $X = a$ or $X = b$, and to be 1 if $a < X < b$, i.e.

$$Z = \tfrac{1}{2}I_{\{a \leqslant X < b\}} + \tfrac{1}{2}I_{\{a < X \leqslant b\}}.$$

Then

$$\psi\{c(X-a)\}-\psi\{c(X-b)\} \to Z \text{ on } \Omega \quad \text{as} \quad c \to \infty.$$

To complete the proof it suffices to show that

$$I(c_n) \to \tilde{F}(b)-\tilde{F}(a) \quad \text{as} \quad n \to \infty \tag{5}$$

whenever $c_1, c_2, \ldots$ are positive numbers satisfying $c_n \to \infty$ as $n \to \infty$.

For any such numbers $c_1, c_2, \ldots$ let

$$X_n = \psi\{c_n(X-a)\}-\psi\{c_n(X-b)\} \qquad (n = 1, 2, \ldots).$$

Then $X_1, X_2, \ldots$ are random variables (why?) and, by (4),

$$|X_n| \leqslant 2k \text{ on } \Omega \qquad (n = 1, 2, \ldots).$$

Since $X_n \to Z$ on Ω as $n \to \infty$, it follows from the bounded convergence theorem that

$$I(c_n) = E(X_n) \to E(Z) \quad \text{as} \quad n \to \infty.$$

Now Z is a simple, non-negative random variable, and so

$$\begin{aligned} E(Z) &= \tfrac{1}{2}\times P(X = a)+\tfrac{1}{2}\times P(X = b)+1\times P(a < X < b) \\ &= \tfrac{1}{2}\{F(a)-F(a-)\}+\tfrac{1}{2}\{F(b)-F(b-)\}+\{F(b-)-F(a)\} \\ &\qquad\qquad\qquad\qquad\qquad\qquad \text{(see Exercise 4.2)} \\ &= \tfrac{1}{2}\{F(b)+F(b-)\}-\tfrac{1}{2}\{F(a)+F(a-)\} \\ &= \tilde{F}(b)-\tilde{F}(a). \end{aligned}$$

Thus (5) holds, and the proof is complete.

It is almost obvious that the inversion formula will suffice to determine the distribution function F, provided the latter has not "too many" points of discontinuity. That this is, in fact, so was shown in Lemma 7.1.

Corollary *Suppose that the random variables X and Y have the same characteristic function. Then X and Y have the same distribution.*

Note. If, conversely, X and Y have the same distribution or, equivalently, the same distribution function, then they have the same characteristic function (see Theorem 5.10).

Proof. Let F and G be the distribution functions of X and Y respectively. It suffices to prove that $F = G$ on R (by Theorem 4.2, corollary 2).

Let

$$\tilde{F}(x) = \tfrac{1}{2}\{F(x-)+F(x)\}, \quad \tilde{G}(x) = \tfrac{1}{2}\{G(x-)+G(x)\} \quad (x \in R).$$

Then, by Theorem 3,

$$\tilde{F}(b)-\tilde{F}(a) = \tilde{G}(b)-\tilde{G}(a) \quad (a, b \in R). \tag{6}$$

Since $0 \leqslant \tilde{F}(a) \leqslant F(a)$ $(a \in R)$, and $F(a) \to 0$ as $a \to -\infty$, it follows that

$$\tilde{F}(a) \to 0 \quad \text{as} \quad a \to -\infty$$

and, similarly, that

$$\tilde{G}(a) \to 0 \quad \text{as} \quad a \to -\infty.$$

Therefore, by (6),

$$\tilde{F}(b) = \tilde{G}(b) \quad (b \in R). \tag{7}$$

Now let x be any real number, and suppose that $b > x$. Then

$$F(x) \leqslant \tilde{F}(b) \leqslant F(b).$$

Since $F(b) \to F(x)$ as $b \to x+$ (because F is everywhere continuous on the right), it follows that

$$\tilde{F}(b) \to F(x) \quad \text{as} \quad b \to x+$$

and, similarly, that

$$\tilde{G}(b) \to G(x) \quad \text{as} \quad b \to x+.$$

Therefore, by (7),

$$F(x) = G(x).$$

This holds for every real number x, and so the proof is complete.

Theorem 4 *Suppose that $X_1, X_2, \ldots, X$ are random variables with corresponding characteristic functions $\varphi_1, \varphi_2, \ldots, \varphi$, and that*

$$X_n \xrightarrow{D} X \quad \textit{as} \quad n \to \infty.$$

Then

$$\varphi_n(t) \to \varphi(t) \quad \textit{as} \quad n \to \infty \quad (t \in R).$$

Proof. In Theorem 7.4 take $f(x)$ to be first $\cos tx$ and then $\sin tx$. It follows immediately that

$$E(\cos tX_n) \to E(\cos tX) \quad \text{as} \quad n \to \infty$$

and

$$E(\sin tX_n) \to E(\sin tX) \quad \text{as} \quad n \to \infty,$$

and so

$$\varphi_n(t) = E(e^{itX_n}) \to E(e^{itX}) = \varphi(t) \quad \text{as} \quad n \to \infty.$$

It is also true that if

$$\varphi_n(t) \to \varphi(t) \quad \text{as} \quad n \to \infty \quad (t \in R)$$

then $X_n \xrightarrow{D} X$ as $n \to \infty$. This is a somewhat deeper result, and a very important one, which we proceed to prove (in a slightly generalised form). The following lemma is a necessary preliminary.

Lemma 2. *Suppose that X is a random variable, and let φ be its characteristic function. Then for every real number $c > 0$*

$$P\left(|X| \geqslant \frac{2}{c}\right) \leqslant \frac{1}{c}\int_{-c}^{c} \{1-\varphi(t)\}\, dt.$$

Note. Since the characteristic function determines the distribution, any probability theory question concerning a random variable can, in theory, be answered when its characteristic function is known. Lemma 2 is an illustration of this principle.

Proof.
$$\frac{1}{c}\int_{-c}^{c} \{1-\varphi(t)\}\, dt = \frac{1}{c}\int_{-c}^{c} E(1-e^{itX})\, dt$$
$$= \frac{1}{c}E\left\{\int_{-c}^{c} (1-e^{itX})\, dt\right\} \quad \text{(by Theorem 5.9C)}.$$

Now when $X \neq 0$

$$\int_{-c}^{c} (1-e^{itX})\, dt = 2c - \frac{e^{icX}-e^{-icX}}{iX} = 2c\left(1-\frac{\sin cX}{cX}\right)$$
$$\geqslant 2c\left(1-\frac{\sin cX}{cX}\right)I_A,$$

where $A = \left\{|X| \geqslant \dfrac{2}{c}\right\}$ (because $(\sin t)/t \leqslant 1$ $(t \in R)$), and

$$\left(1-\frac{\sin cX}{cX}\right)I_A \geqslant \tfrac{1}{2}I_A,$$

because
$$1-\frac{\sin cX}{cX} \geqslant \tfrac{1}{2} \quad \text{if} \quad |X| \geqslant \frac{2}{c}.$$

Therefore, if $X \neq 0$,

$$\int_{-c}^{c} (1-e^{itX})\,dt \geqslant cI_A,$$

and this inequality also holds when $X = 0$. Consequently

$$\frac{1}{c}\int_{-c}^{c} \{1-\varphi(t)\}\,dt \geqslant \frac{1}{c}E(cI_A) = P(A),$$

which is the desired result.

Theorem 5 *Suppose that $X_1, X_2, \ldots$ are random variables with corresponding characteristic functions $\varphi_1, \varphi_2, \ldots$. Suppose also that*

$$\varphi_n(t) \to \textit{a limit } \varphi(t) \quad \textit{as} \quad n \to \infty \quad (t \in R), \tag{8}$$

where φ is continuous at 0. Then φ is the characteristic function of some random variable X, and

$$X_n \xrightarrow{D} X \quad \textit{as} \quad n \to \infty.$$

Proof. We shall need to use Theorem 7.5, and so must first establish the condition of that theorem.

Let ε be any positive number. Since

$$\varphi(0) = \lim_{n\to\infty} \varphi_n(0) = 1,$$

by (8) and Theorem 1 (i), and since φ is continuous at 0, there exists a positive number c such that

$$|\varphi(t)-1| < \tfrac{1}{2}\varepsilon \, (-c \leqslant t \leqslant c). \tag{9}$$

Since $|\varphi_n(t)| \leqslant 1$ (all t) (by Theorem 1 (ii)), the bounded convergence theorem for Lebesgue integrals may be applied to give

$$\int_{-c}^{c} \varphi_n(t)\,dt \to \int_{-c}^{c} \varphi(t)\,dt \quad \text{as} \quad n \to \infty. \tag{10}$$

Here the integral on the right-hand side must be interpreted in the Lebesgue sense, for though the real and imaginary parts of $\varphi = \lim_{n\to\infty} \varphi_n$ are certainly Borel measurable, we do not yet know that φ is continuous (except at 0). However, with a little ingenuity, it is possible to avoid the explicit use of Lebesgue theory (see the note on page 145).

Now

$$\left|\frac{1}{c}\int_{-c}^{c} \{1-\varphi(t)\}\,dt\right| \leqslant \frac{1}{c}\int_{-c}^{c} |1-\varphi(t)|\,dt < \varepsilon$$

by (9), and so, by (10),

$$\left|\frac{1}{c}\int_{-c}^{c}\{1-\varphi_n(t)\}\,dt\right| < \varepsilon$$

for all sufficiently large n. Therefore, by Lemma 2,

$$P\left(|X_n| \geqslant \frac{2}{c}\right) < \varepsilon$$

for all sufficiently large n, and so, for such n,

$$F_n\left(\frac{2}{c}\right)-F_n\left(-\frac{2}{c}\right) = P\left(-\frac{2}{c} < X_n \leqslant \frac{2}{c}\right) \geqslant P\left(|X_n| < \frac{2}{c}\right) > 1-\varepsilon,$$

where F_n is the distribution function of X_n.

By Theorem 7.5, there exists a subsequence $F_{n_1}, F_{n_2}, \ldots$ of $F_1, F_2, \ldots$ and a distribution function F such that

$$F_{n_p} \xrightarrow{D} F \quad \text{as} \quad p \to \infty,$$

i.e.

$$X_{n_p} \xrightarrow{D} X \quad \text{as} \quad p \to \infty, \tag{11}$$

where X is any random variable having the distribution function F.

Let φ^* be the characteristic function of X. Then, by Theorem 4,

$$\varphi_{n_p}(t) \to \varphi^*(t) \quad \text{as} \quad p \to \infty \quad (t \in R),$$

and so, by (8),

$$\varphi^*(t) = \varphi(t) \quad (t \in R).$$

Therefore φ is the characteristic function of X.

It remains to prove that

$$X_n \xrightarrow{D} X \quad \text{as} \quad n \to \infty$$

(as opposed to the weaker statement (11)).

If this is not so, there exists a real number b such that F is continuous at b and

$$F_n(b) \nrightarrow F(b) \quad \text{as} \quad n \to \infty.$$

Since the sequence $F_1(b), F_2(b), \ldots$ is bounded, it contains a subsequence $F_{m_1}(b), F_{m_2}(b), \ldots$ convergent to some limit $k \neq F(b)$. By Theorem 7.5, there exists a subsequence $F_{m_{r_1}}, F_{m_{r_2}}, \ldots$ of $F_{m_1}, F_{m_2}, \ldots$ and a distribution function G such that

$$F_{m_{r_s}} \xrightarrow{D} G \quad \text{as} \quad s \to \infty.$$

Let ψ be the characteristic function of G (i.e. of a random variable having the distribution function G). Then, for all $t \in R$,

$$\psi(t) = \lim_{s\to\infty} \varphi_{m_{r_s}}(t) \quad \text{(by Theorem 4)}$$
$$= \varphi(t) \quad \text{(by (8))}.$$

Therefore, by Theorem 3, corollary,

$$G(x) = F(x) \quad (x \in R),$$

and so $F_{m_{r_s}} \xrightarrow{D} F$ as $s \to \infty$. But F is continuous at b, and so

$$F(b) = \lim_{s\to\infty} F_{m_{r_s}}(b) = \lim_{r\to\infty} F_{m_r}(b) = k \neq F(b).$$

This contradiction completes the proof of the theorem.

Note. As stated above, the explicit use of Lebesgue theory in the proof of Theorem 5 may be avoided.

Let H be the distribution function defined by

$$H(t) = \begin{cases} 0 & \text{if} \quad t \leqslant -c \\ \frac{1}{2}\left(1+\dfrac{t}{c}\right) & \text{if} \quad -c \leqslant t \leqslant c \\ 1 & \text{if} \quad t \geqslant c, \end{cases}$$

and let Q be the corresponding probability measure on $\mathscr{B}$ (see Theorem 4.3). Then $\varphi_1, \varphi_2, \ldots$ and $\varphi = \lim_{n\to\infty} \varphi_n$ may be regarded as (complex-valued) random variables defined on the probability space $(R, \mathscr{B}, Q)$, and, by the bounded convergence theorem (Theorem 5.6, corollary),

$$E(1-\varphi_n) \to E(1-\varphi) \quad \text{as} \quad n \to \infty.$$

Now
$$E(1-\varphi_n) = \int_{-\infty}^{\infty} (1-\varphi_n)\, dH \quad \text{(by Theorem 5.11)}$$
$$= \int_{-c}^{c} (1-\varphi_n)\, dH$$
$$= \frac{1}{2c}\int_{-c}^{c} (1-\varphi_n) \qquad \text{(a Riemann integral)}.$$

Also, if $J(t) = 1$ if $-c \leqslant t \leqslant c$, and $= 0$ otherwise,

$$|E(1-\varphi)| \leqslant E(|1-\varphi|)$$
$$= E(|1-\varphi|J) \quad \text{(by Exercise 5.5 (b))}$$
$$< E(\tfrac{1}{2}\varepsilon J) \quad \text{(by (9) and Exercise 5.2 (ii))}$$
$$= E(\tfrac{1}{2}\varepsilon) \quad \text{(by Exercise 5.5 (b))}$$
$$= \tfrac{1}{2}\varepsilon.$$

Therefore

$$\left|\frac{1}{c}\int_{-c}^{c}(1-\varphi_n)\right| < \varepsilon$$

for all sufficiently large n, and the proof is completed as before.

Corollary *Suppose that $X_1, X_2, \ldots, X$ are random variables with corresponding characteristic functions $\varphi_1, \varphi_2, \ldots, \varphi$. Then*

$$X_n \xrightarrow{D} X \quad as \quad n \to \infty$$

if and only if

$$\varphi_n(t) \to \varphi(t) \quad as \quad n \to \infty \quad (t \in R).$$

Proof. Theorems 4 and 5.

It is now possible to give the proof of the weak law of large numbers which we promised earlier.

Theorem 6 (*the weak law of large numbers*) *Suppose that $X_1, X_2, \ldots$ are independent observations of a random variable X with finite expected value μ. Then*

$$\bar{X}_n = \frac{1}{n}(X_1+X_2+ \ldots +X_n) \xrightarrow{P} \mu \quad as \quad n \to \infty.$$

Proof. Let φ be the characteristic function of X. Then, since $X_1, X_2, \ldots$ have the same distribution as X, they also have the same characteristic function φ. Therefore, by Theorem 2 (ii), $X_1+X_2+ \ldots +X_n$ has the characteristic function $\varphi^n(t)$, and so, by Exercise 2, $\bar{X}_n$ has the characteristic function $\varphi^n(t/n)$.

Now, by Theorem 1 (v),

$$\varphi(t) = 1+it\{\mu+o(1)\} \quad \text{as} \quad t \to 0.$$

Therefore, for each real number t,

$$\varphi^n\left(\frac{t}{n}\right) = \left\{1+\frac{it}{n}(\mu+\varepsilon_n)\right\}^n,$$

where $\varepsilon_n \to 0$ as $n \to \infty$, and so

$$\varphi^n\left(\frac{t}{n}\right) \to e^{i\mu t} \quad \text{as} \quad n \to \infty.$$

The limit function $e^{i\mu t}$ is the characteristic function of the constant random variable μ, and so, by Theorem 5, corollary,

$$\bar{X}_n \xrightarrow{D} \mu \quad \text{as} \quad n \to \infty.$$

The result now follows from Theorem 7.3.

We come finally to the central limit theorem. This tells us, roughly speaking, that if $X_1, X_2, \ldots$ are independent observations of a random variable X with a finite second moment then, for large n, $X_1+X_2+\ldots+X_n$ is approximately normally distributed with mean $nE(X)$ and variance $n \operatorname{var} X$. More precisely, it tells us that if

$$Y_n = \{X_1+X_2+\ldots+X_n-nE(X)\}/\sqrt{(n \operatorname{var} X)} \qquad (n = 1, 2, \ldots)$$

then the sequence $\{Y_n\}$ of random variables converges in distribution to a standardised normal random variable (i.e. a normal (0, 1) random variable). (Note that Y_n has been obtained from $X_1+X_2+\ldots+X_n$ by a change of origin and scale in such a way as to have zero mean and unit variance.)

It is clear that the very generality of this result ensures that the normal distribution plays a key role in probability theory, and that it would still play such a role even if it did not have its many applications in the sciences.

If, in particular, we assume that $X_1, X_2, \ldots$ are independent observations of a binomial $(1, p)$ random variable X, then $E(X) = p$, $\operatorname{var} X = pq$, where $q = 1-p$, and $S_n = X_1+X_2+\ldots+X_n$ is a binomial (n, p) random variable. Since the distribution function Φ of a normal (0, 1) random variable is given by

$$\Phi(x) = (2\pi)^{-\frac{1}{2}} \int_{-\infty}^{x} \exp\left(-\tfrac{1}{2}t^2\right) dt \qquad (x \in R),$$

the central limit theorem tells us that

$$P\left(\frac{S_n-np}{\sqrt{(npq)}} \leqslant x\right) \to \Phi(x) \quad \text{as} \quad n \to \infty \qquad (x \in R).$$

It follows that if x and y are any two real numbers with $x < y$ then

$$P\left(x < \frac{S_n-np}{\sqrt{(npq)}} \leqslant y\right)$$

is approximately $\Phi(y)-\Phi(x)$ if n is large. This is the "normal approximation to the binomial" with which the reader is probably already familiar.

Theorem 7 (*the central limit theorem*) *Suppose that $X_1, X_2, \ldots$ are independent observations of a random variable X with finite second moment, and let*

$$\mu = E(X), \qquad \sigma = \sqrt{(\operatorname{var} X)},$$

where it is assumed that $\sigma > 0$. Then, for every real number x,

$$P\left(\frac{X_1+X_2+\ldots+X_n-n\mu}{\sigma\sqrt{n}} \leqslant x\right) \to \Phi(x) \quad \text{as} \quad n \to \infty, \tag{12}$$

where

$$\Phi(x) = (2\pi)^{-\frac{1}{2}} \int_{-\infty}^{x} \exp(-\tfrac{1}{2}t^2)\, dt.$$

Note. If $\sigma = 0$, each of $X_1, X_2, \ldots$ is a.s. equal to μ.

Proof. Let F_n and φ_n denote respectively the distribution function and the characteristic function of

$$(X_1 + X_2 + \ldots + X_n - n\mu)/(\sigma\sqrt{n}) \qquad (n = 1, 2, \ldots).$$

Then we have to show that

$$F_n \xrightarrow{D} \Phi \quad \text{as} \quad n \to \infty$$

(which is equivalent to (12) since Φ is everywhere continuous).

Since Φ is the distribution function of a normal (0, 1) random variable, and the characteristic function of such a random variable is $\exp(-\frac{1}{2}t^2)$ (see Exercise 4), we have to show that

$$\varphi_n(t) \to \exp(-\tfrac{1}{2}t^2) \quad \text{as} \quad n \to \infty \quad (t \in R).$$

Let $\varphi(t)$ denote the common characteristic function of $X_1 - \mu, X_2 - \mu, \ldots$. (It simplifies the algebra to define φ in this way, rather than as the common characteristic function of $X_1, X_2, \ldots$.) Since $X_1 - \mu, X_2 - \mu, \ldots$ are independent (see Exercise 6.7(i)), the characteristic function of

$$X_1 + X_2 + \ldots + X_n - n\mu = \sum_{k=1}^{n} (X_k - \mu)$$

is $\varphi^n(t)$, and consequently that of

$$(X_1 + X_2 + \ldots + X_n - n\mu)/(\sigma\sqrt{n}),$$

i.e. $\varphi_n(t)$, is $\varphi^n\{t/(\sigma\sqrt{n})\}$ (by Exercise 2).

By Theorem 1 (v)

$$\varphi(t) = 1 - \tfrac{1}{2}\sigma^2 t^2\{1 + o(1)\} \quad \text{as} \quad t \to 0,$$

since $E(X - \mu) = 0$ and $E\{(X - \mu)^2\} = \sigma^2$. Therefore, for each fixed real number t,

$$\varphi^n\left(\frac{t}{\sigma\sqrt{n}}\right) = \left\{1 - \frac{t^2}{2n}(1 + \varepsilon_n)\right\}^n,$$

where $\varepsilon_n \to 0$ as $n \to \infty$, and so

$$\varphi^n\left(\frac{t}{\sigma\sqrt{n}}\right) \to \exp(-\tfrac{1}{2}t^2) \quad \text{as} \quad n \to \infty.$$

This completes the proof of the central limit theorem.

It is possible to extend the central limit theorem to deal with random variables $X_1, X_2, \ldots$ which, though independent, are not identically distributed (see Exercise 19).

Exercises

1. Suppose that the random variable X is a constant c on Ω. Prove that the characteristic function of X is e^{ict}.

2. Suppose that the random variable X has the characteristic function $\varphi(t)$. Prove that $aX+b$ has the characteristic function $e^{ibt}\varphi(at)$.

3. Verify the following results:

(i) The characteristic function of a binomial (n, p) random variable is $(1-p+pe^{it})^n$.

(ii) The characteristic function of a Poisson (μ) random variable is $\exp\{\mu(e^{it}-1)\}$.

4. Suppose that a random variable has one of the probability distributions of Exercise 5.23. Verify that the corresponding characteristic function is

(i) $(e^{ibt}-e^{iat})/\{i(b-a)t\}$;
(ii) $\lambda/(\lambda-it)$;
(iii) $\exp(i\mu t-\frac{1}{2}\sigma^2t^2)$;
(iv) $e^{-|t|}$.

5. For any characteristic function φ prove that

$$\varphi(-t) = \overline{\varphi(t)} \quad (t \in R).$$

6. Let φ be the characteristic function of a random variable X. Prove that φ is uniformly continuous on R.

Hint. Note that

$$|\varphi(t+h)-\varphi(t)| \leqslant E(|e^{ihX}-1|),$$

and prove that the right-hand side $\to 0$ as $h \to 0$.

7. Suppose that X is a random variable with characteristic function φ, and that $\varphi''(0)$ is finite. Prove that $E(X^2)$ is finite.

Hint. $\varphi''(0) = \lim_{h\to 0} \{\varphi(h)+\varphi(-h)-2\varphi(0)\}/h^2$,

and so

$$-\varphi''(0) = \lim_{n\to\infty} E\{n^2 \sin^2(X/n)\}.$$

Apply Fatou's lemma (Exercise 5.15).

8. Suppose that $\varphi(t) = \exp(-c\,|t|^k)$, where $c > 0$ and $k > 0$, is the characteristic function of a random variable X. Prove that $k \leqslant 2$.

Hint. If $k > 2$, then $\varphi''(0) = 0$, and so $E(X^2) < \infty$ (by Exercise 7). It follows successively that (i) $E(X^2) = 0$, (ii) $X = 0$ a.s., (iii) $\varphi(t) = 1$ $(t \in R)$.

9. Suppose that the random variables $X_1, X_2, \ldots, X_n$ are independent, and that X_k is normal (μ_k, σ_k^2) $(k = 1, 2, \ldots, n)$. Determine the distribution of $X_1+X_2+\ldots+X_n$.

Hint. Find the characteristic function of $X_1+X_2+\ldots+X_n$.

10. Suppose that $X_1, X_2, \ldots, X_n$ are independent random variables having the Cauchy distribution (see Exercise 5.23 (iv)). What is the distribution of

$$\frac{1}{n}(X_1+X_2+\ldots+X_n)?$$

11. Suppose that the random variable X has the characteristic function φ. Prove that φ is real if and only if the distribution of X is symmetric about 0 (i.e. if and only if X and $-X$ have the same distribution).

Hint. The characteristic function of $-X$ is $\varphi(-t)$; use Exercise 5.

12. Suppose that φ is a characteristic function. Prove that φ^2 and $|\varphi|^2$ are characteristic functions.

Hint. Suppose that φ is the characteristic function of a random variable X, and let X_1 and X_2 be independent observations of X (the question of existence involved here was discussed on page 122). Consider the characteristic functions of X_1+X_2 and X_1-X_2.

13. Suppose that X_1, X_2 are independent observations of a random variable X. Prove that X_1+X_2 has the same distribution as X if and only if $X = 0$ a.s.

Hint. Show that the characteristic function of X must be 1 throughout some neighbourhood of the origin. Now refer to the hint for Exercise 8.

14. Suppose that the random variable X has characteristic function φ. Prove that

$$P(X = x) = \lim_{c\to\infty} \frac{1}{2c}\int_{-c}^{c} e^{-ixt}\varphi(t)\,dt \quad (x \in R).$$

Hint. Show that, for $c > 0$,

$$\frac{1}{2c}\int_{-c}^{c} e^{-ixt}\varphi(t)\,dt = E\left\{\frac{1}{c}\int_{0}^{c} \cos t(X-x)\,dt\right\}$$
$$= E[\psi\{c(X-x)\}],$$

where $\psi(t) = (\sin t)/t$ if $t \neq 0$, and $= 1$ if $t = 0$. Note that $\psi\{c(X-x)\} \to I_{\{X=x\}}$ as $c \to \infty$, and apply the bounded convergence theorem.

15. Suppose that the random variable X has characteristic function φ, and that the Cauchy–Riemann integral

$$\int_{-\infty}^{\infty} \varphi(t)\, dt$$

is absolutely convergent. Prove that X has the continuous probability density function f given by

$$f(x) = \frac{1}{2\pi}\int_{-\infty}^{\infty} e^{-ixt}\varphi(t)\, dt \quad (x \in R).$$

Hints. (i) F is continuous on R (by Exercise 14).

(ii) Suppose that $\varepsilon > 0$, and choose $c > 0$ so that

$$\int_{-\infty}^{-c} |\varphi(t)|\, dt < \varepsilon \quad \text{and} \quad \int_{c}^{\infty} |\varphi(t)|\, dt < \varepsilon.$$

Then, if x and $x+h \in R$,

$$|f(x+h)-f(x)| < \frac{|h|}{2\pi}\int_{-c}^{c} |t\varphi(t)|\, dt + \varepsilon,$$

from which it follows that f is continuous on R.

(iii) $$\frac{F(x+h)-F(x)}{h} = \frac{1}{2\pi}\int_{-\infty}^{\infty} \frac{1-e^{-iht}}{iht} e^{-ixt}\varphi(t)\, dt \quad \text{(by Theorem 3)}$$

$$\to f(x) \quad \text{as} \quad h \to 0$$

by a similar argument to that employed in (ii). Thus

$$F'(x) = f(x) \quad \text{and} \quad F(x) = \int_{-\infty}^{x} f(t)\, dt \quad (x \in R).$$

Finally refer to Exercise 5.25.

16. Suppose that φ_n is the characteristic function of a normal $(0, n)$ random variable $(n = 1, 2, \ldots)$. Prove that $\varphi_n(t)$ tends to a limit $\varphi(t)$ as $n \to \infty$ $(t \in R)$, but that φ is not continuous at the origin.

If $F_1, F_2, \ldots$ are the distribution functions corresponding to $\varphi_1, \varphi_2, \ldots$ respectively, determine $\lim_{n\to\infty} F_n(x)$.

17. (i) Prove that $\prod_{k=1}^{n} \cos(t/2^k) \to \dfrac{\sin t}{t}$ as $n \to \infty$.

(ii) The random variables $X_1, X_2, \ldots$ are independent, and X_n takes the values $+2^{-n}$ and -2^{-n} each with probability $\frac{1}{2}$ $(n = 1, 2, \ldots)$. Prove that

$$X_1+X_2+\ldots+X_n \xrightarrow{D} X \quad \text{as} \quad n \to \infty,$$

where X has the uniform distribution on $(-1, 1)$.

18. (i) Suppose that $X_1, X_2, \ldots$ are independent Poisson (1) random variables. Show, by using the results of this chapter or otherwise, that $Y_n = X_1+X_2+\ldots+X_n$ is a Poisson (n) random variable ($n = 1, 2, \ldots$).

(ii) Show that $(Y_n-n)/\sqrt{n} \xrightarrow{D} Z$ as $n \to \infty$, where Z is normal (0, 1).

(iii) Deduce that

$$e^{-n} \sum_{k=0}^{n} n^k/k! \to \tfrac{1}{2} \quad \text{as} \quad n \to \infty.$$

The following exercise outlines a proof of a useful extension of the central limit theorem due to Liapounov. The condition that $X_1, X_2, \ldots$ be identically distributed is dropped, but an additional condition is imposed that ensures that each of the random variables makes only a relatively small contribution to the sum $\sum_{k=1}^{n} \{X_k - E(X_k)\}$ for large n. This additional condition is formalsed in (∗) below, and requires that the X's have finite third moments.

19. Suppose that the random variables $X_1, X_2, \ldots$ are independent and have finite third moments. For $n = 1, 2, \ldots$ let

$$\mu_n = E(X_n), \quad \sigma_n = (\text{var } X_n)^{\frac{1}{2}}, \quad \varrho_n = E(|X_n-\mu_n|^3)\}^{\frac{1}{3}},$$

$$m_n = \mu_1+\mu_2+\ldots+\mu_n,$$

$$s_n = (\sigma_1^2+\sigma_2^2+\ldots+\sigma_n^2)^{\frac{1}{2}},$$

and

$$r_n = (\varrho_1^3+\varrho_2^3+\ldots+\varrho_n^3)^{\frac{1}{3}},$$

and suppose that $s_n > 0$ ($n = 1, 2, \ldots$) and

$$r_n/s_n \to 0 \quad \text{as} \quad n \to \infty. \tag{∗}$$

Prove that

$$P\left(\frac{X_1+X_2+\ldots+X_n-m_n}{s_n} \leqslant x\right) \to \Phi(x) \quad \text{as} \quad n \to \infty \quad (x \in R),$$

where $\Phi(x)$ is defined as for Theorem 7.

Outline of proof. For $n = 1, 2, \ldots$ let φ_n and ψ_n be the characteristic functions of $X_n - \mu_n$ and $(X_1 + X_2 + \ldots + X_n - m_n)/s_n$ respectively, and so

$$\psi_n(t) = \prod_{k=1}^{n} \varphi_k(t/s_n) \quad (t \in R).$$

Now let t be a fixed real number, and establish the following results:

(i) $|\varphi_k(t) - 1 + \frac{1}{2}\sigma_k^2 t^2| \leqslant \frac{1}{6}|t|^3 \varrho_k^3 \qquad (k = 1, 2, \ldots).$

Hint. Use Lemma 1.

(ii) $|\log(1+z) - z| \leqslant |z|^2 \quad (|z| \leqslant \frac{1}{2}),$

where the logarithm has its principal value.

(iii) For all sufficiently large n, and $k = 1, 2, \ldots, n$,

$$\begin{aligned}
&\frac{1}{2}\frac{\sigma_k^2}{s_n^2}t^2 + \frac{1}{6}\frac{\varrho_k^3}{s_n^3}|t|^3 \\
&\quad \leqslant \frac{1}{2}\frac{\varrho_k^2}{s_n^2}t^2 + \frac{1}{6}\frac{\varrho_k^3}{s_n^3}|t|^3 \quad \text{(by Exercise 5.4)} \\
&\quad \leqslant \left(\frac{\varrho_k}{s_n}\right)^{\frac{3}{2}} \{\tfrac{1}{2}t^2 + \tfrac{1}{6}|t|^3\} \quad \text{(since } \varrho_k/s_n \leqslant r_n/s_n < 1 \text{ for all sufficiently large } n\text{)} \\
&\quad < \tfrac{1}{2}.
\end{aligned}$$

(iv) For all sufficiently large n, and $k = 1, 2, \ldots, n$,

$$\left|\log \varphi_k\left(\frac{t}{s_n}\right) + \frac{1}{2}\frac{\sigma_k^2}{s_n^2}t^2\right| \leqslant \frac{\varrho_k^3}{s_n^3}\{\tfrac{1}{6}|t|^3 + (\tfrac{1}{2}t^2 + \tfrac{1}{6}|t|^3)^2\}.$$

(v) For all sufficiently large n

$$|\log \psi_n(t) + \tfrac{1}{2}t^2| \leqslant \left(\frac{r_n}{s_n}\right)^3 \{\tfrac{1}{6}|t|^3 + (\tfrac{1}{2}t^2 + \tfrac{1}{6}|t|^3)^2\}.$$

(vi) $\psi_n(t) \to \exp(-\frac{1}{2}t^2)$ as $n \to \infty$.

20. Let $X_1, X_2, \ldots$ be independent random variables for which

$$P(X_n = n) = P(X_n = -n) = \tfrac{1}{2} \qquad (n = 1, 2, \ldots).$$

Show that

$$(3/n^3)^{\frac{1}{2}}(X_1 + X_2 + \ldots + X_n) \xrightarrow{D} Z \quad \text{as} \quad n \to \infty,$$

where Z is normal (0, 1).

Hint. Use Exercises 19 and 7.19.

Appendix A

Lebesgue theory

It has been necessary on occasion to use some results relating to the Lebesgue integral, either because they play an essential part in a proof (as with that of Theorem 8.5) or because they enable us to attain the appropriate degree of generality (as with the discussion of absolutely continuous distributions towards the end of Chapter 5). Moreover, certain probability spaces are most naturally described in terms of Lebesgue measure. If, for example, we consider the experiment of choosing a number at random in the interval $(0, 1)$, a suitable probability space $(\Omega, \mathscr{F}, P)$ is obtained by taking Ω to be $(0, 1)$, $\mathscr{F}$ to be the class of all Lebesgue measurable subsets of $(0, 1)$, and P to be Lebesgue measure on $\mathscr{F}$.

Lebesgue measure on the real line R arises out of the attempt to extend the idea of length from the class of finite intervals to a much more extensive class of sets – the class $\mathscr{L}$ of *Lebesgue measurable sets*. More precisely, the theory establishes the existence of a sigma-algebra $\mathscr{L}$ in R and a function m: $\mathscr{L} \to [0, \infty]$ with the following properties:

(a) $\mathscr{L}$ contains all the intervals (finite or infinite);
(b) if I is a finite interval then

$$m(I) = \text{the length of } I,$$

i.e. $m\{(a, b)\} = m\{(a, b]\} = m\{[a, b)\} = m\{[a, b]\} = b-a$ for any real numbers a, b with $a < b$;

(c) if $A_1, A_2, \ldots$ are disjoint sets of $\mathscr{L}$ then

$$m\left(\bigcup_1^\infty A_n\right) = \sum_1^\infty m(A_n);$$

(d) if $B \in \mathscr{L}$, $m(B) = 0$, and $A \subseteq B$, then $A \in \mathscr{L}$.

If A is a set of $\mathscr{L}$, $m(A)$ is called the *Lebesgue measure* of A.

Since $\mathscr{L}$ contains, in particular, all the open intervals, it contains all the open sets in R, as any open set in R is a countable union of open intervals. Therefore, by definition of $\mathscr{B}$,

$$\mathscr{L} \supseteq \mathscr{B}. \tag{1}$$

The restriction m_1 of m to $\mathscr{B}$ satisfies the analogues of (b) and (c) above, but not that of (d). Because the analogue of (d) does not hold for the measure space $(R, \mathscr{B}, m_1)$, this measure space is not complete. In fact, $(R, \mathscr{L}, m)$ is obtained from $(R, \mathscr{B}, m_1)$ by the process of completion described in Theorem 2.5.

It is easy to show that the Lebesgue measure of any one-point set is zero, and it then follows from (c) that the Lebesgue measure of any countable set is zero. For example, if Q denotes the set of all rational numbers, then $m(Q) = 0$. It is also easy to show that if I is any infinite interval then $m(I) = \infty$.

A function f: $R \to R$ is *Lebesgue measurable* if it is $\mathscr{L}$-measurable, i.e. if

$$f^{-1}\{(-\infty, c]\} \in \mathscr{L}$$

for every real number c (see the definition of measurability in Chapter 3). This is equivalent to saying that

$$f^{-1}(\mathscr{B}) \subseteq \mathscr{L}$$

(see Theorem 3.1). The results of Theorems 3.3–3.7 hold for Lebesgue measurable functions. Further, if f is Borel measurable then it is Lebesgue measurable (by (1) above); in particular, if f is continuous or monotonic on R, then it is Lebesgue measurable.

More generally, if the domain of definition of the real-valued function f is some subset D of R, and $A \subseteq D$, then f is said to be *Lebesgue measurable on A* if

$$A \cap f^{-1}\{(-\infty, c]\} \in \mathscr{L}$$

for every real number c, or, equivalently, if

$$A \cap f^{-1}(B) \in \mathscr{L} \quad (B \in \mathscr{B}). \tag{2}$$

It follows from (2), on taking $B = R$, that $A \in \mathscr{L}$.

The results of Chapter 3 can be extended in an obvious manner to functions which are Lebesgue measurable on A. In particular, if the set A is Lebesgue measurable, and f is continuous or monotonic on A, then f is Lebesgue measurable on A. Note also that if f is Lebesgue measurable on A, and A_1 is a Lebesgue measurable subset of A, then f is Lebesgue measurable on A_1.

Now suppose that f is Lebesgue measurable on A. Then the Lebesgue integral $\int_A f$ can be defined in three stages, paralleling the corresponding definition of $E(X)$ in Chapter 5.

Firstly, suppose that f is non-negative and simple on A (to say that f is simple on A means that it takes only a finite number of distinct values on A). Then there exist non-negative real numbers

$$a_1, a_2, \ldots, a_k$$

and disjoint Lebesgue measurable sets

$$A_1, A_2, \ldots, A_k$$

whose union is A such that

$$f = a_r \text{ on } A_r \qquad (r = 1, 2, \ldots, k).$$

Then $\int_A f$ is defined to be

$$\sum_{r=1}^{k} a_r m(A_r).$$

Any product $0 \times \infty$ which occurs in the above sum is given the value 0.

It can be shown, as in Chapter 5, that $\int_A f$ is uniquely determined by the above definition, and that the results corresponding to Lemma 1 of that chapter continue to hold. It can also be shown that the result corresponding to Lemma 2 of that chapter holds, but the proof is now a little more difficult because of the possible occurrence of sets of infinite measure.

Secondly, suppose that f is non-negative on A. Then, as in Theorem 3.7, there exist functions $f_1, f_2, \ldots$ which are Lebesgue measurable and simple on A and satisfy

$$0 \leqslant f_1 \leqslant f_2 \leqslant \ldots \text{ on } A$$

and

$$f_n \to f \text{ on } A \quad \text{as} \quad n \to \infty.$$

Then $\int_A f$ is now defined to be

$$\lim_{n\to\infty} \int_A f_n .$$

As in Chapter 5, it can be shown that $\int_A f$ is uniquely determined by this definition, and that the results corresponding to Lemma 3 of that chapter continue to hold.

Finally, if f can take values of either sign on A, then $\int_A f$ is said to *exist* if $\int_A f^+$ and $\int_A f^-$ are not both infinite, and then

$$\int_A f = \int_A f^+ - \int_A f^-.$$

In particular, f is said to be *integrable in the Lebesgue sense over A* if $\int_A f$ is finite, i.e. if $\int_A f^+$ and $\int_A f^-$ are both finite.

It can be shown that if $[a, b]$ is a closed finite interval and f is integrable in the Riemann sense over $[a, b]$ then f is integrable in the Lebesgue sense over $[a, b]$ and

$$\int_{[a, b]} f = \int_a^b f(x)\, dx.$$

Other notations for the Lebesgue integral $\int_A f$ are $\int_A f\, dm$ (which emphasises that the measure used is the Lebesgue measure m) and $\int_A f(x)\, dx$. If, in particular, $A = [a, b]$, then $\int_A f(x)\, dx$ is often written $\int_a^b f(x)\, dx$. It will usually be clear from the context whether this denotes the Lebesgue or the Riemann integral, but even if it is not, the ambiguity is immaterial as the two integrals then have the same value.

With two exceptions the obvious analogues of Theorems 1–6 of Chapter 5, with $E(X)$ replaced by $\int_A f$, continue to hold. The two exceptions are Theorem 4, corollary, and Theorem 6, corollary (the bounded convergence theorem). These now require the additional condition that $m(A)$ be finite. In particular, if $f_1, f_2, \ldots$ are measurable on A, and f_n tends to a finite limit f on A as $n \to \infty$, then

$$\int_A f_n \to \int_A f \quad \text{as} \quad n \to \infty$$

if either

$$0 \leqslant f_1 \leqslant f_2 \leqslant \ldots \text{ on } A$$

(the monotone convergence theorem), or there exists a function g which is integrable in the Lebesgue sense over A and is such that

$$|f_n| \leqslant g \text{ on } A \qquad (n = 1, 2, \ldots)$$

(the dominated convergence theorem).

A further result, which can be easily established, is that if f is Lebesgue measurable and non-negative on $A \cup B$, where A and B are disjoint and Lebesgue measurable, then

$$\int_{A \cup B} f = \int_A f + \int_B f.$$

It is hoped that this appendix contains a summary of the ideas and results of Lebesgue theory which is adequate for an understanding of the references to it in this book. The reader who would like to learn more about this theory is referred to one of the many available texts. A particularly full and clear treatment is given in H. Kestelman, *Modern Theories of Integration* (Dover, 1960), though the integral is there defined in a different way from that given above.

Appendix B

The extension theorem

Proofs of the extension theorem (Theorem 4.2) can be found in P. R. Halmos, *Measure Theory* (Van Nostrand, 1950) (see §§ 10–13) and in J. F. C. Kingman and S. J. Taylor, *Introduction to Measure and Probability* (Cambridge University Press, 1966) (see §§ 3.1, 3.3 and 4.1). However, in these books the extension theorem is proved under rather more general conditions than we need here, and it is possible to simplify the proof somewhat when the measure of the whole space is known to be finite, as is the case with probability theory, since $P(\Omega) = 1$. Furthermore, it will be convenient to have the extension theorem in the precise form in which we shall require it.

Although the proof is somewhat lengthy, it can be broken down into a number of relatively simple steps, and many of these will be left to the reader. Thus much of this appendix can be regarded as a collection of easy exercises.

We start with a statement of the extension theorem.

The extension theorem

Let $\mathscr{S}$ be a semi-algebra of sets in Ω, and let $\mu\colon \mathscr{S} \to R$ satisfy the following conditions:

(a) $\mu(A) \geqslant 0$ *for every* $A \in \mathscr{S}$;

(b) $\mu(\Omega) = 1$;

(c) *if* $A_1, A_2, \ldots$ *are disjoint and* $\in \mathscr{S}$, *and also* $\bigcup_1^\infty A_n \in \mathscr{S}$, *then*

$$\mu\left(\bigcup_1^\infty A_n\right) = \sum_1^\infty \mu(A_n).$$

Then there exists one and only one probability measure P on $\sigma(\mathscr{S})$ which is equal to μ on $\mathscr{S}$, i.e. for which

$$P(A) = \mu(A) \quad \textit{for every} \quad A \in \mathscr{S}.$$

The following lemma lists the properties of μ which we shall require.

Lemma 1. (i) $\mu(\emptyset) = 0$.

(ii) *If* $A_1, A_2, \ldots, A_n$ *are disjoint and* $\in \mathscr{S}$, *and also* $\bigcup_1^n A_i \in \mathscr{S}$, *then*

$$\mu\left(\bigcup_1^n A_i\right) = \sum_1^n \mu(A_i).$$

(iii) *If* $A_1, A_2, \ldots, A_n$ *are disjoint and* $\in \mathscr{S}$, $A \in \mathscr{S}$, *and* $\bigcup_1^n A_i \subseteq A$, *then*

$$\sum_1^n \mu(A_i) \leqslant \mu(A).$$

In particular, if $A, B \in \mathscr{S}$, *and* $A \subseteq B$, *then*

$$\mu(A) \leqslant \mu(B).$$

(iv) $\mu(A) \leqslant 1$ $(A \in \mathscr{S})$.

(v) *If* $A_1, A_2, \ldots, A_n \in \mathscr{S}$, *and* $\bigcup_1^n A_i \in \mathscr{S}$, *then*

$$\mu\left(\bigcup_1^n A_i\right) \leqslant \sum_1^n \mu(A_i).$$

(vi) *If* $A_1, A_2, \ldots, A_n$ *and* $A \in \mathscr{S}$, *and* $A \subseteq \bigcup_1^n A_i$, *then*

$$\mu(A) \leqslant \sum_1^n \mu(A_i).$$

(vii) *If* $A_1, A_2, \ldots \in \mathscr{S}$, *and* $\bigcup_1^\infty A_n \in \mathscr{S}$, *then*

$$\mu\left(\bigcup_1^\infty A_n\right) \leqslant \sum_1^\infty \mu(A_n).$$

(viii) *If* $A_1, A_2, \ldots$ *and* $A \in \mathscr{S}$, *and* $A \subseteq \bigcup_1^\infty A_n$, *then*

$$\mu(A) \leqslant \sum_1^\infty \mu(A_n).$$

Note. The reader will observe that to prove parts (i)–(vi) it suffices to assume only that μ satisfies conditions (a), (b) and the "finite" form of (c), viz. if $A_1, A_2, \ldots, A_n$ are disjoint and $\in \mathscr{S}$, and also $\bigcup_1^n A_i \in \mathscr{S}$, then

$$\mu\left(\bigcup_1^n A_i\right) = \sum_1^n \mu(A_i)$$

(i.e. instead of assuming that μ is *countably additive* on $\mathscr{S}$, it suffices to assume only that it is *finitely additive* on $\mathscr{S}$). These conditions are easily seen to be satisfied by the function $\mu : \mathscr{I} \to R$ of Theorem 4.3, and the results of Lemma 1 (iii) and (vi) were, in fact, used in the proof of that theorem.

We now define a function $\mu^* : \mathscr{P}(\Omega) \to R$ which is an extension of μ: $\mathscr{S} \to R$. The restriction of μ^* to $\sigma(\mathscr{S})$ will be shown to be the desired probability measure P.

For any set A in Ω let $\mu^*(A)$ be the greatest lower bound of the set of numbers

$$\left\{\sum_1^\infty \mu(B_n) : B_1, B_2, \ldots \in \mathscr{S} \text{ and } A \subseteq \bigcup_1^\infty B_n\right\}.$$

Clearly $\mu^*(A)$ is defined for every set A in Ω and

$$0 \leqslant \mu^*(A) \leqslant 1 \quad (A \subseteq \Omega). \tag{1}$$

Lemma 2. (i) $\mu^*(\emptyset) = 0$.

(ii) *If* $A \subseteq B \subseteq \Omega$, *then* $\mu^*(A) \leqslant \mu^*(B)$.

(iii) *For any sets* $A_1, A_2, \ldots \subseteq \Omega$,

$$\mu^*\left(\bigcup_1^\infty A_n\right) \leqslant \sum_1^\infty \mu^*(A_n).$$

(iv) *For any sets* $A_1, A_2, \ldots, A_n \subseteq \Omega$,

$$\mu^*\left(\bigcup_1^n A_i\right) \leqslant \sum_1^n \mu^*(A_i).$$

(v) $\mu^* = \mu$ *on* $\mathscr{S}$.

Note. Parts (i)–(iii) assert that μ^* is an *outer measure* on $\mathscr{P}(\Omega)$.

Hints. (iii) Suppose that $\varepsilon > 0$. For each $n = 1, 2, \ldots$ there exist sets B_{n1}, $B_{n2}, \ldots \in \mathscr{S}$ such that

$$A_n \subseteq \bigcup_{p=1}^\infty B_{np} \quad \text{and} \quad \sum_{p=1}^\infty \mu(B_{np}) < \mu^*(A_n) + \varepsilon 2^{-n}.$$

(v) Suppose that $A \in \mathscr{S}$. Clearly $\mu^*(A) \leqslant \mu(A)$. To obtain the reverse inequality, use Lemma 1 (viii).

Next we define a set A in Ω to be *measurable* if

$$\mu^*(C) = \mu^*(C \cap A) + \mu^*(C \cap A^c) \tag{2}$$

for *every* $C \subseteq \Omega$. In view of Lemma 2 (iv), (2) may be replaced by

$$\mu^*(C) \geqslant \mu^*(C \cap A) + \mu^*(C \cap A^c).$$

It is not easy to give any simple motivation of this definition (see Kingman and Taylor, p. 75). However, the reader familiar with Lebesgue measure will know that if Lebesgue measurability of a set in R is defined in terms of its inner and outer measure, then it is measurable in the sense of such a definition if and only if it is measurable in the sense of the definition we have just given.

The class of all measurable sets will be denoted by $\mathscr{M}$.

Lemma 3. (i) $\Omega \in \mathscr{M}$.

(ii) *If* $A \in \mathscr{M}$, *then* $A^c \in \mathscr{M}$.
(iii) *If* $A, B \in \mathscr{M}$, *then* $A \cap B \in \mathscr{M}$.
(iv) *If* $A_1, A_2, \ldots$ *are disjoint and* $\in \mathscr{M}$, *then*

$$A = \bigcup_1^\infty A_n \in \mathscr{M}$$

and

$$\mu^*(A) = \sum_1^\infty \mu^*(A_n). \tag{3}$$

(v) *If* $A_1, A_2, \ldots \in \mathscr{M}$, *then* $\bigcup_1^\infty A_n \in \mathscr{M}$.

(vi) $\mathscr{M}$ *is a sigma-algebra of sets in* Ω.

Proof

(iii) For all $C \subseteq \Omega$

$$\begin{aligned}\mu^*(C) &= \mu^*(C \cap A) + \mu^*(C \cap A^c) \quad \text{(because } A \in \mathscr{M}\text{)}\\ &= \mu^*(C \cap A \cap B) + \mu^*(C \cap A \cap B^c) + \mu^*(C \cap A^c)\\ &\qquad\qquad\qquad\qquad \text{(because } B \in \mathscr{M}\text{)}\\ &\geqslant \mu^*(C \cap A \cap B) + \mu^*\{C \cap (A \cap B)^c\}\end{aligned}$$

(by Lemma 2 (iv), because $(C \cap A \cap B^c) \cup (C \cap A^c) = C \cap (A \cap B)^c$).

(iv) For all $C \subseteq \Omega$

$$\begin{aligned}\mu^*(C) &= \mu^*(C \cap A_1) + \mu^*(C \cap A_1^c) \quad \text{(because } A_1 \in \mathscr{M}\text{)}\\ &= \mu^*(C \cap A_1) + \mu^*(C \cap A_2) + \mu^*(C \cap A_1^c \cap A_2^c)\\ &\qquad\qquad \text{(because } A_2 \in \mathscr{M}\text{, and } A_1 \cap A_2 = \emptyset\text{)},\end{aligned}$$

and so on. Thus, for all n,

$$\begin{aligned}\mu^*(C) &= \sum_1^n \mu^*(C \cap A_i) + \mu^*\left\{C \cap \left(\bigcup_1^n A_i\right)^c\right\}\\ &\geqslant \sum_1^n \mu^*(C \cap A_i) + \mu^*(C \cap A^c) \quad \text{(by Lemma 2 (ii))}.\end{aligned}$$

On letting n tend to infinity, it follows that

$$\mu^*(C) \geqslant \sum_1^\infty \mu^*(C \cap A_n) + \mu^*(C \cap A^c)$$
$$\geqslant \mu^*(C \cap A) + \mu^*(C \cap A^c) \quad \text{(by Lemma 2 (iii))}.$$

Therefore $A \in \mathscr{M}$, and (3) follows on putting $C = A$.

(v) Note that

$$\bigcup_1^\infty A_n = A_1 \cup (A_1^c \cap A_2) \cup (A_1^c \cap A_2^c \cap A_3) \cup \ldots .$$

Lemma 4. $\mathscr{M} \supseteq \mathscr{S}$.

Proof. Suppose that $A \in \mathscr{S}$ and $C \subseteq \Omega$.

Let $B_1, B_2, \ldots$ be any sets of $\mathscr{S}$ for which

$$C \subseteq \bigcup_1^\infty B_n.$$

Then $C \cap A \subseteq \bigcup_1^\infty (B_n \cap A)$, and so

$$\mu^*(C \cap A) \leqslant \sum_1^\infty \mu(B_n \cap A).$$

Also, for each n, $B_n \cap A^c$ can be expressed as the union of a finite number of disjoint sets of $\mathscr{S}$, say $B_{n1}, B_{n2}, \ldots, B_{nk_n}$. Therefore

$$C \cap A^c \subseteq \bigcup_{n=1}^\infty \bigcup_{p=1}^{k_n} B_{np},$$

and so

$$\mu^*(C \cap A^c) \leqslant \sum_{n=1}^\infty \sum_{p=1}^{k_n} \mu(B_{np}).$$

Therefore

$$\mu^*(C \cap A) + \mu^*(C \cap A^c)$$
$$\leqslant \sum_{n=1}^\infty \left\{\mu(B_n \cap A) + \sum_{p=1}^{k_n} \mu(B_{np})\right\}$$
$$= \sum_{n=1}^\infty \mu(B_n).$$

Therefore

$$\mu^*(C \cap A) + \mu^*(C \cap A^c) \leqslant \mu^*(C), \quad \text{and so} \quad A \in \mathscr{M}.$$

It has been shown that $\mathcal{M}$ is a sigma-algebra (Lemma 3 (vi)) containing all the sets of $\mathcal{S}$ (Lemma 4), and that the restriction of μ^* to $\mathcal{M}$ is a probability measure on $\mathcal{M}$ (see (1), Lemma 2 (v) and Lemma 3 (iv)). Since $\mathcal{M} \supseteq \sigma(\mathcal{S})$, the restriction P of μ^* to $\sigma(\mathcal{S})$ is a probability measure on $\sigma(\mathcal{S})$. Also $P = \mu$ on $\mathcal{S}$ (by Lemma 2 (v)).

It remains to prove that P is unique.

Let P_1 be any probability measure on $\sigma(\mathcal{S})$ which is equal to μ on $\mathcal{S}$. It will be shown that, for any $A \in \sigma(\mathcal{S})$,

$$P_1(A) \leqslant P(A). \tag{4}$$

This will hold with A^c (which also belongs to $\sigma(\mathcal{S})$) in place of A, and so

$$P_1(A^c) \leqslant P(A^c).$$

But

$$P_1(A)+P_1(A^c) = P(A)+P(A^c) = 1$$

(because P and P_1 are probability measures on $\sigma(\mathcal{S})$), and so

$$P_1(A) = P(A) \quad (A \in \sigma(\mathcal{S})).$$

Thus the proof of the extension theorem will be completed once (4) has established for any $A \in \sigma(\mathcal{S})$.

Suppose then that $A \in \sigma(\mathcal{S})$, and let $B_1, B_2, \ldots$ be any sets of $\mathcal{S}$ for which $A \subseteq \bigcup_1^\infty B_n$. Then

$$\begin{aligned} P_1(A) &\leqslant P_1\left(\bigcup_1^\infty B_n\right) && \text{(by Lemma 2.2 (v))} \\ &\leqslant \sum_1^\infty P_1(B_n) && \text{(by Theorem 2.1)} \\ &= \sum_1^\infty \mu(B_n) && \text{(because } P_1 = \mu \text{ on } \mathcal{S}\text{).} \end{aligned}$$

Therefore

$$\begin{aligned} P_1(A) &\leqslant \mu^*(A) && \text{(by definition of } \mu^*\text{)} \\ &= P(A) && \text{(by definition of } P\text{),} \end{aligned}$$

and (4) is established.

It can be shown that, denoting the restriction of μ^* to $\mathcal{M}$ by $\bar{P}$, the probability space $(\Omega, \mathcal{M}, \bar{P})$ is complete and is, in fact, the completion of $(\Omega, \sigma(\mathcal{S}), P)$.

Appendix C

The Riemann-Stieltjes integral

This appendix contains a brief treatment of the Riemann–Stieltjes integral for those who are unfamiliar with that topic. We shall concentrate on those theorems which we require in this book, and shall not strive after the most general form of their statements. When proofs are straightforward they will be left to the reader, and so once again much of the appendix can be regarded as a collection of easy exercises.

Suppose that f and g are two real-valued functions defined on a closed interval $[a, b]$. Let P be a *partition* of $[a, b]$, i.e. let P be a set of real numbers $\{x_0, x_1, x_2, \ldots, x_n\}$ for which n is a positive integer and

$$a = x_0 < x_1 < x_2 < \ldots < x_n = b.$$

Let

$$N(P) = \max\{x_r - x_{r-1}: \quad r = 1, 2, \ldots, n\}$$

be the *norm* of the partition P.

Definition. f is *integrable in the Riemann–Stieltjes sense with respect to g over* $[a, b]$, or, more briefly, $\int_a^b f\,dg$ *exists*, if there exists a number I such that to every number $\varepsilon > 0$ there corresponds a number $\delta > 0$ such that

$$\left| \sum_{r=1}^{n} f(t_r)\{g(x_r) - g(x_{r-1})\} - I \right| < \varepsilon \tag{1}$$

for every partition P of $[a, b]$ for which $N(P) < \delta$ and, for each such partition, every choice of numbers $t_1, t_2, \ldots, t_n$, for which

$$x_{r-1} \leqslant t_r \leqslant x_r \qquad (r = 1, 2, \ldots, n).$$

This will be more briefly and less formally expressed in the following way:

$$\sum_{r=1}^{n} f(t_r)\{g(x_r) - g(x_{r-1})\} \to I \quad \text{as} \quad N(P) \to 0.$$

If such a number I exists it is clearly unique, and we write

$$\int_a^b f(x)\,dg(x) = I \quad \text{or} \quad \int_a^b f\,dg = I.$$

In particular, if $g(x) = x\ (a \leqslant x \leqslant b)$, then f is integrable in the Riemann–Stieltjes sense with respect to g over $[a, b]$ if and only if f is integrable in the Riemann sense over $[a, b]$, and the Riemann–Stieltjes integral $\left(\int_a^b f\,dg\right)$ and the Riemann integral $\left(\int_a^b f\right)$ are equal.

In Theorems 1–8, all functions will be assumed to be real-valued and defined on $[a, b]$.

Theorem 1

(i) *If* $f = k$ *(a constant) on* $[a, b]$, *then*

$$\int_a^b f\,dg \text{ (exists and) } = k\{g(b)-g(a)\} \quad (\text{all } g).$$

(ii) *If* g *is constant on* $[a, b]$, *then*

$$\int_a^b f\,dg \text{ (exists and) } = 0 \quad (\text{all } f).$$

(iii) *If* $\int_a^b f\,dg$ *exists and* k *is a constant, then* $\int_a^b kf\,dg$ *and* $\int_a^b f\,d(kg)$ *exist and have the common value* $k\int_a^b f\,dg$.

(iv) *If* $\int_a^b f_1\,dg$ *and* $\int_a^b f_2\,dg$ *exist, then*

$$\int_a^b (f_1+f_2)\,dg \text{ (exists and) } = \int_a^b f_1\,dg + \int_a^b f_2\,dg.$$

(v) *If* $\int_a^b f\,dg_1$ *and* $\int_a^b f\,dg_2$ *exist, then*

$$\int_a^b f\,d(g_1+g_2) \text{ (exists and) } = \int_a^b f\,dg_1 + \int_a^b f\,dg_2.$$

Proof. The proof is trivial and is left to the reader.

Theorem 2 *Suppose that*

(i) $\int_a^b f_1 \, dg$ *and* $\int_a^b f_2 \, dg$ *exist,*
(ii) $f_1 \leqslant f_2$ *on* $[a, b]$,
(iii) g *is non-decreasing on* $[a, b]$.

Then
$$\int_a^b f_1 \, dg \leqslant \int_a^b f_2 \, dg.$$

Proof. Once again the proof is trivial.

Theorem 3 (*integration by parts*)

Suppose that $\int_a^b f \, dg$ *exists. Then* $\int_a^b g \, df$ *exists and*

$$\int_a^b g \, df = f(b)\,g(b) - f(a)\,g(a) - \int_a^b f \, dg.$$

Proof. Let $P = \{x_0, x_1, \ldots, x_n\}$ be a partition of $[a, b]$, and suppose that

$$x_{r-1} \leqslant t_r \leqslant x_r \qquad (r = 1, 2, \ldots, n).$$

Then

$$\sum_{r=1}^{n} g(t_r)\{f(x_r) - f(x_{r-1})\} = f(b)\,g(b) - f(a)\,g(a) - S, \tag{2}$$

where

$$S = f(a)\{g(t_1) - g(a)\} + \sum_{r=1}^{n-1} f(x_r)\{g(t_{r+1}) - g(t_r)\} + f(b)\{g(b) - g(t_n)\}. \tag{3}$$

Now $a \leqslant t_1 \leqslant t_2 \leqslant \ldots \leqslant t_n \leqslant b$, and so the set of numbers $\{a, t_1, t_2, \ldots, t_n, b\}$ gives a partition Q, say, of $[a, b]$ if repetitions are ignored. (If, for example, $t_1 = a$, this number is counted only once in the partition, and the first sub-interval of the partition is $[a, t_2]$. Note that, correspondingly, the first term on the right-hand side of (3) vanishes.) Since

$$a \leqslant a \leqslant t_1 \leqslant x_1 \leqslant t_2 \leqslant \ldots \leqslant t_{n-1} \leqslant x_{n-1} \leqslant t_n \leqslant b \leqslant b,$$

the sum S is an approximating sum for $\int_a^b f \, dg$ based on the partition Q.

Now

$$\begin{aligned} N(Q) &= \max\,(t_1 - a,\ t_2 - t_1,\ \ldots,\ t_n - t_{n-1},\ b - t_n) \\ &\leqslant 2N(P), \end{aligned}$$

and so, by the definition of $\int_a^b f\,dg$,

$$S \to \int_a^b f\,dg \quad \text{as} \quad N(P) \to 0.$$

The result follows from (2).

Now let us assume further that f is bounded on $[a, b]$ and g is non-decreasing on $[a, b]$. Let

$$M = \sup\{f(x) : a \leqslant x \leqslant b\} \quad \text{and} \quad m = \inf\{f(x) : a \leqslant x \leqslant b\}.$$

Further, if $P = \{x_0, x_1, \ldots, x_n\}$ is a partition of $[a, b]$ let

$$M_r = M_r(P) = \sup\{f(x) : x_{r-1} \leqslant x \leqslant x_r\}$$

and

$$m_r = m_r(P) = \inf\{f(x) : x_{r-1} \leqslant x \leqslant x_r\}$$

for $r = 1, 2, \ldots, n$. Let also

$$S(P) = \sum_{r=1}^{n} M_r\{g(x_r)-g(x_{r-1})\}$$

and

$$s(P) = \sum_{r=1}^{n} m_r\{g(x_r)-g(x_{r-1})\}.$$

Then

$$m\{g(b)-g(a)\} \leqslant s(P) \leqslant S(P) \leqslant M\{g(b)-g(a)\}. \tag{4}$$

Finally let

$$J = \inf\{S(P) \colon P \text{ is a partition of } [a, b]\}$$

and

$$j = \sup\{s(P) \colon P \text{ is a partition of } [a, b]\}.$$

It follows from (4) that J and j are finite.

Lemma 1. *Suppose that f is bounded on $[a, b]$ and g is non-decreasing on $[a, b]$. Then*

(i) *for any partitions P_1, P_2 of $[a, b]$ for which $P_1 \subseteq P_2$*

$$s(P_1) \leqslant s(P_2) \leqslant S(P_2) \leqslant S(P_1);$$

(ii) *for any partitions P_1, P_2 of $[a, b]$*

$$s(P_1) \leqslant S(P_2);$$

(iii) $j \leqslant J$.

Proof. The proof is virtually identical with that of the corresponding result for Riemann integrals, and is left to the reader.

Corollary *For every partition P of $[a, b]$*

$$s(P) \leqslant j \leqslant J \leqslant S(P). \tag{5}$$

Lemma 2. *Suppose that f is bounded on $[a, b]$ and g is non-decreasing on $[a, b]$. Then $\int_a^b f\,dg$ exists if and only if to every number $\varepsilon > 0$ there corresponds a number $\delta > 0$ such that*

$$S(P) - s(P) < \varepsilon \quad (N(P) < \delta).$$

Outline of proof. To prove the "only if" part, note that it follows from (1) that

$$I - \varepsilon \leqslant s(P) \leqslant S(P) \leqslant I + \varepsilon \quad (N(P) < \delta).$$

To prove the "if" part, note that, by (5), the given condition implies that $J = j = I$, say, and then, again by (5),

$$I - \varepsilon < s(P) \leqslant \sum_{r=1}^{n} f(t_r)\{g(x_r) - g(x_{r-1})\} \leqslant S(P) < I + \varepsilon$$

whenever $N(P) < \delta$.

Theorem 4 *Suppose that f is bounded on $[a, b]$ and g is non-decreasing on $[a, b]$. Suppose also that $a < c < b$. Then the following results hold:*

(i) *If $\int_a^b f\,dg$ exists, then $\int_a^c f\,dg$, $\int_c^b f\,dg$ exist and*

$$\int_a^b f\,dg = \int_a^c f\,dg + \int_c^b f\,dg. \tag{6}$$

(ii) *If $\int_a^c f\,dg$, $\int_c^b f\,dg$ exist and f is continuous at c, then $\int_a^b f\,dg$ exists and (6) holds.*

Proof. (i) The existence of the integrals is a consequence of Lemma 2, and then (6) follows immediately.

(ii) Let $P = \{x_0, x_1, \ldots, x_n\}$ be a partition of $[a, b]$, and suppose that $x_{k-1} \leqslant c \leqslant x_k$. Then if $x_{r-1} \leqslant t_r \leqslant x_r$ $(r = 1, 2, \ldots, n)$

$$\sum_{r=1}^{n} f(t_r)\{g(x_r)-g(x_{r-1})\}$$
$$= \left[\sum_{r=1}^{k-1} f(t_r)\{g(x_r)-g(x_{r-1})\}+f(c)\{g(c)-g(x_{k-1})\}\right]$$
$$+\left[f(c)\{g(x_k)-g(c)\}+\sum_{r=k+1}^{n} f(t_r)\{g(x_r)-g(x_{r-1})\}\right]$$
$$+\{f(t_k)-f(c)\}\{g(x_k)-g(x_{k-1})\}$$
$$\to \int_a^c f\,dg+\int_c^b f\,dg+0 \quad \text{as} \quad N(P)\to 0.$$

Note. It is not in general true that if $\int_a^c f\,dg$ and $\int_c^b f\,dg$ exist then $\int_a^b f\,dg$ exists. For let

$$f(x) = g(x) = 0 \quad (0 \leqslant x < 1),$$
$$f(1) = 1, \quad g(1) = 0,$$

and

$$f(x) = g(x) = 1 \quad (1 < x \leqslant 2).$$

Then the reader will easily verify that $\int_0^1 f\,dg$ and $\int_1^2 f\,dg$ exist, but $\int_0^2 f\,dg$ does not exist.

The following theorem gives an important sufficient condition for the existence of $\int_a^b f\,dg$.

Theorem 5 *Suppose that f is continuous on* $[a, b]$ *and g is non-decreasing on* $[a, b]$. *Then* $\int_a^b f\,dg$ *and* $\int_a^b g\,df$ *exist.*

Proof. Suppose that $\varepsilon > 0$. Since f is continuous on $[a, b]$ it is uniformly continuous on $[a, b]$, and so there exists a number $\delta > 0$ such that

$$\max\{M_r - m_r : r = 1, 2, \ldots, n\} < \varepsilon$$

whenever $N(P) < \delta$. Therefore

$$S(P)-s(P) \leqslant \varepsilon\{g(b)-g(a)\} \quad (N(P) < \delta),$$

and the existence of $\int_a^b f\,dg$ follows from Lemma 2.

The existence of $\int_a^b g\,df$ now follows from Theorem 3.

Corollary 1 *Suppose that f is continuous on $[a, b]$ and g is of bounded variation on $[a, b]$. Then $\int_a^b f\,dg$ and $\int_a^b g\,df$ exist.*

Proof. There exist functions g_1, g_2 which are non-decreasing on $[a, b]$ and are such that $g = g_1 - g_2$ on $[a, b]$; now apply Theorem 1 (iii) and (v).

Corollary 2 *Suppose that f is continuous on $[a, b]$ and g is non-decreasing on $[a, b]$. Then*

$$\left|\int_a^b f\,dg\right| \leqslant \int_a^b |f|\,dg.$$

Proof. Since both integrals exist, and $-|f| \leqslant f \leqslant |f|$ on $[a, b]$, the result follows from Theorem 2.

The evaluation of Riemann–Stieltjes integrals

Theorem 6 *Suppose that*

(i) $a = y_0 < y_1 < y_2 < \ldots < y_m < b$,
(ii) *g is constant in each of the intervals*

$$[y_{r-1}, y_r) \qquad (r = 1, 2, \ldots, m) \quad \textit{and} \quad [y_m, b],$$

(iii) *f is continuous at $y_1, y_2, \ldots, y_m$.*

Then

$$\int_a^b f\,dg \text{ (exists and) } = \sum_{r=1}^{m} f(y_r)\{g(y_r) - g(y_{r-1})\}.$$

Proof. Left to the reader.

Theorem 7 *Suppose that*

(i) *f is either continuous on $[a, b]$ or of bounded variation on $[a, b]$,*
(ii) *g' is continuous on $[a, b]$.*

Then

$$\int_a^b f\,dg = \int_a^b f(t)\,g'(t)\,dt.$$

Notes (a). The integral on the right-hand side is a Riemann integral.

(b) Since g has a continuous derivative g' on $[a, b]$, g is both continuous and of bounded variation on $[a, b]$, and so $\int_a^b f\,dg$ exists (by Theorem 5, corollary 1).

Proof. Let $P = \{x_0, x_1, \ldots, x_n\}$ be a partition of $[a, b]$. For $r = 1, 2, \ldots, n$ there exists a number $t_r \in (x_{r-1}, x_r)$ such that

$$g(x_r) - g(x_{r-1}) = g'(t_r)(x_r - x_{r-1}).$$

Therefore

$$\sum_{r=1}^{n} f(t_r)\{g(x_r) - g(x_{r-1})\} = \sum_{r=1}^{n} f(t_r) g'(t_r)(x_r - x_{r-1}).$$

As $N(P) \to 0$, the left-hand side tends to $\int_a^b f\, dg$ (which is known to exist), and the right-hand side tends to $\int_a^b f(t)\, g'(t)\, dt$.

Corollary *Suppose that*

(i) f *is continuous on* $[a, b]$,
(ii) g *is non-decreasing and continuous on* $[a, b]$ *and its derivative* g' *is continuous on* (a, b).

Then

$$\int_a^b f\, dg = \int_a^b f(t) g'(t)\, dt. \tag{7}$$

Note. The integral on the right-hand side is a Cauchy–Riemann integral. Its convergence will be established in the course of the proof.

Proof. Suppose that $a < a' < b' < b$. Then

$$\int_{a'}^{b'} f\, dg = \int_{a'}^{b'} f(t)\, g'(t)\, dt$$

(by the theorem). Now let

$$M = \sup\{|f(x)| : a \leqslant x \leqslant b\}.$$

Then

$$\begin{aligned} \left| \int_{a'}^{b'} f\, dg - \int_a^b f\, dg \right| &= \left| \int_a^{a'} f\, dg + \int_{b'}^b f\, dg \right| \quad \text{(by Theorem 4)} \\ &\leqslant \int_a^{a'} |f|\, dg + \int_{b'}^b |f|\, dg \\ &\leqslant M\{g(a') - g(a)\} + M\{g(b) - g(b')\} \\ &\to 0 \quad \text{as} \quad a' \to a \quad \text{and} \quad b' \to b. \end{aligned}$$

Thus the Cauchy–Riemann integral is convergent, and (7) holds.

Theorems 6 and 7 show how certain finite sums and Riemann integrals can be expressed in a common notational form $\int_a^b f\,dg$. This is particularly useful when g is the distribution function of a random variable X, for it enables us to express $E(X)$ as a Riemann–Stieltjes integral (with, admittedly, infinite limits), and give a unified treatment of its properties, whether the distribution of X be discrete (compare Theorem 6) or absolutely continuous (compare Theorem 7) or, indeed, neither. This was touched upon in Chapter 5.

Theorem 8 *Suppose that*

(i) f *is continuous on* $[a, b]$,
(ii) $g_1, g_2, \ldots$ *and* g *are non-decreasing on* $[a, b]$,
(iii) *there exists a subset* C *of* $[a, b]$ *which is everywhere dense in* $[a, b]$, *contains both* a *and* b, *and is such that*

$$g_n(x) \to g(x) \quad \textit{as} \quad n \to \infty \quad (x \in C).$$

Then

$$\int_a^b f\,dg_n \to \int_a^b f\,dg \quad \textit{as} \quad n \to \infty.$$

Note. The theorem will be applied in the particular case when C is the set of those points of (a, b) at which g is continuous, together with a and b. That C is everywhere dense in $[a, b]$ then follows from Lemma 7.1.

Proof. Suppose that $\varepsilon > 0$. Since f is continuous on $[a, b]$ it is uniformly continuous there, and so there exists a number $\delta > 0$ such that

$$|f(x) - f(x')| < \varepsilon$$

for all $x, x' \in [a, b]$ for which $|x - x'| < \delta$.

It is possible to choose a finite number of points of C, say $x_0, x_1, \ldots, x_p$, such that

$$a = x_0 < x_1 < \ldots < x_p = b$$

and

$$|x_r - x_{r-1}| < \delta \qquad (r = 1, 2, \ldots, p).$$

Then

$$\left| \int_a^b f\,dg - \sum_{r=1}^{p} f(x_r)\{g(x_r) - g(x_{r-1})\} \right|$$
$$= \left| \sum_{r=1}^{p} \int_{x_{r-1}}^{x_r} \{f(x) - f(x_r)\}\,dg(x) \right|$$

$$\leqslant \sum_{r=1}^{p} \int_{x_{r-1}}^{x_r} |f(x)-f(x_r)| \, dg(x) \quad \text{(by Theorem 5, corollary 2)}$$

$$\leqslant \varepsilon \sum_{r=1}^{p} \{g(x_r)-g(x_{r-1})\}$$

$$= \varepsilon\{g(b)-g(a)\}.$$

Similarly, for all n,

$$\left| \int_a^b f \, dg_n - \sum_{r=1}^{p} f(x_r)\{g_n(x_r)-g_n(x_{r-1})\} \right| \leqslant \varepsilon\{g_n(b)-g_n(a)\}.$$

Therefore

$$\left| \int_a^b f \, dg_n - \int_a^b f \, dg \right|$$

$$\leqslant \left| \sum_{r=1}^{p} f(x_r)\{g_n(x_r)-g_n(x_{r-1})\} - \sum_{r=1}^{p} f(x_r)\{g(x_r)-g(x_{r-1})\} \right|$$

$$+\varepsilon\{g(b)-g(a)\}+\varepsilon\{g_n(b)-g_n(a)\}.$$

Now let n tend to infinity. Then, by (iii), the right-hand side tends to $2\varepsilon\{g(b)-g(a)\}$, and so

$$\left| \int_a^b f \, dg_n - \int_a^b f \, dg \right| < \varepsilon[2\{g(b)-g(a)\}+1]$$

for all sufficiently large n. The result follows.

Infinite Riemann–Stieltjes integrals

Let a be a real number, and let J denote the infinite interval $[a, \infty)$. Suppose that

(i) f and g are real-valued functions defined on J,

(ii) f is continuous on J,

(iii) g is non-decreasing on J.

It follows that $\int_a^b f \, dg$ exists for all real numbers b with $b > a$.

Definition. The infinite Riemann–Stieltjes integral $\int_a^\infty f\,dg$ is *convergent* if $\int_a^b f\,dg$ tends to a limit as $b \to \infty$, and its *value* is then

$$\lim_{b \to \infty} \int_a^b f\,dg,$$

which will also be denoted by $\int_a^\infty f\,dg$. The integral $\int_a^\infty f\,dg$ is *absolutely convergent* if $\int_a^\infty |f|\,dg$ is convergent.

Suppose in particular that $f \geqslant 0$ on J. Then, for $a < b < c$,

$$\int_a^c f\,dg - \int_a^b f\,dg = \int_b^c f\,dg \quad \text{(by Theorem 4)}$$
$$\geqslant 0.$$

It follows that $\int_a^\infty f\,dg$ is convergent if and only if the set of numbers $\left\{\int_a^b f\,dg : b > a\right\}$ is bounded above.

More generally, i.e. if f is not assumed to be non-negative on J, $\int_a^\infty f\,dg$ is absolutely convergent if and only if the set of numbers $\left\{\int_a^b |f|\,dg : b > a\right\}$ is bounded above.

It follows from this last remark that $\int_a^\infty f\,dg$ is absolutely convergent if f and g are bounded on J.

Theorem 9 *Suppose that f is continuous on J, g is non-decreasing on J, and $\int_a^\infty f\,dg$ is absolutely convergent. Then $\int_a^\infty f\,dg$ is convergent.*

Proof. Since, by Theorem 2,

$$\int_a^b f^+\,dg, \quad \int_a^b f^-\,dg \leqslant \int_a^b |f|\,dg \quad (b > a),$$

it follows from what has been stated above that $\int_a^\infty f^+\,dg$ and $\int_a^\infty f^-\,dg$ are convergent. The result follows.

Corresponding definitions and results hold for infinite Riemann–Stieltjes integrals of the form $\int_{-\infty}^{a} f\,dg$.

Suppose now that f and g are real-valued functions defined on R, f is continuous on R, and g is non-decreasing on R.

Definition. The infinite Riemann–Stieltjes integral $\int_{-\infty}^{\infty} f\,dg$ is *convergent* if, for some real a, $\int_{-\infty}^{a} f\,dg$ and $\int_{a}^{\infty} f\,dg$ are convergent, and its *value* is then

$$\int_{-\infty}^{a} f\,dg + \int_{a}^{\infty} f\,dg,$$

which will also be denoted by $\int_{-\infty}^{\infty} f\,dg$.

Whether $\int_{-\infty}^{\infty} f\,dg$ is convergent, and its value when it is convergent, are clearly independent of a.

Absolute convergence is defined as before, and the analogue of Theorem 9 holds, i.e. if $\int_{-\infty}^{\infty} f\,dg$ is absolutely convergent then it is convergent.

Appendix D

The strong law of large numbers

The meaning and significance of the strong law were fully discussed in Chapter 7, and so we can now concentrate on its proof.

We start with some preliminary lemmas.

Lemma 1. *Suppose that* $X_1, X_2, \ldots, Y_1, Y_2, \ldots$ *are random variables, and that*

$$\sum_{n=1}^{\infty} P(X_n \neq Y_n) \quad \textit{is convergent.}$$

Then

$$\frac{1}{n} \sum_{k=1}^{n} (X_k - Y_k) \to 0 \quad \textit{a.s as} \quad n \to \infty.$$

Proof. Let $A_n = \{X_n \neq Y_n\}$ $(n = 1, 2, \ldots)$, and so $\sum_{n=1}^{\infty} P(A_n)$ is convergent. From the Borel–Cantelli lemmas it follows that, with probability one, only a finite number of the events $A_1, A_2, \ldots$ occur, or, equivalently, that $X_n = Y_n$ for all sufficiently large n. Finally, note that if

$$X_n(\omega) = Y_n(\omega) \quad \text{for all sufficiently large } n$$

then, clearly,

$$\frac{1}{n} \sum_{k=1}^{n} \{X_k(\omega) - Y_k(\omega)\} \to 0 \quad \text{as} \quad n \to \infty.$$

Lemma 2. (*Kolmogorov's inequality*). *Suppose that* $X_1, X_2, \ldots, X_n$ *are independent random variables with zero means and finite second moments, and let*

$$v_k = \text{var}\, X_k = E(X_k^2) \qquad (k = 1, 2, \ldots, n).$$

Then, for any $\varepsilon > 0$,

$$P(\max\{|X_1+X_2+\ldots+X_k| : k = 1, 2, \ldots, n\} \geqslant \varepsilon) \leqslant \varepsilon^{-2} \sum_{k=1}^{n} v_k.$$

Note. Chebyshev's inequality (Exercise 5.8), is a particular case ($n = 1$) of Kolmogorov's inequality.

Proof. For $k = 1, 2, \ldots, n$ let

$$S_k = X_1+X_2+\ldots+X_k$$

and

$$A_k = \{|S_k| \geqslant \varepsilon\} \cap \bigcap_{i=1}^{k-1} \{|S_i| < \varepsilon\}$$

(in particular, $A_1 = \{|S_1| \geqslant \varepsilon\}$). Then the sets $A_1, A_2, \ldots, A_n$ are disjoint and, if

$$A = \bigcup_{k=1}^{n} A_k,$$

we have to prove that

$$P(A) \leqslant \varepsilon^{-2} \sum_{k=1}^{n} v_k.$$

Now

$$E(S_n^2) \geqslant E(S_n^2 I_A) = \sum_{k=1}^{n} E(S_n^2 I_{A_k}).$$

For $k = 1, 2, \ldots, n-1$

$$E(S_n^2 I_{A_k}) = E\left\{S_k^2 I_{A_k} + 2 \sum_{i=k+1}^{n} S_k X_i I_{A_k} + (S_n - S_k)^2 I_{A_k}\right\}$$

$$\geqslant E(S_k^2 I_{A_k}) + 2 \sum_{i=k+1}^{n} E(X_i S_k I_{A_k}).$$

For $i = k+1, k+2, \ldots, n$, X_i and $S_k I_{A_k}$ are independent; this follows from Theorem 6.1 (b) since I_{A_k}, and therefore $S_k I_{A_k}$, is a Borel measurable function of $X_1, X_2, \ldots, X_k$, and, furthermore, the random variables $X_1, X_2, \ldots, X_k$, and, X_i are indepepdent. Since $E(X_i) = 0$ and $E(S_k I_{A_k})$ is finite (see Exercise 5.3), it now follows from Theorem 6.3 that

$$E(X_i S_k I_{A_k}) = E(X_i)\,E(S_k I_{A_k}) = 0.$$

Therefore

$$E(S_n^2 I_{A_k}) \geqslant E(S_k^2 I_{A_k})$$

for $k = 1, 2, \ldots, n-1$ and, trivially, for $k = n$.

Now

$$\begin{aligned} E(S_n^2) &\geqslant \sum_{k=1}^{n} E(S_n^2 I_{A_k}) \\ &\geqslant \sum_{k=1}^{n} E(S_k^2 I_{A_k}) \\ &\geqslant \sum_{k=1}^{n} \varepsilon^2 E(I_{A_k}) \quad (\text{since } |S_k| \geqslant \varepsilon \text{ on } A_k \ (k = 1, 2, \ldots, n)) \\ &= \varepsilon^2 \sum_{k=1}^{n} P(A_k) = \varepsilon^2 P(A). \end{aligned}$$

The result follows since

$$E(S_n^2) = \operatorname{var} S_n = \sum_{k=1}^{n} v_k.$$

Lemma 3. *Suppose that $X_1, X_2, \ldots$ are independent random variables with zero means and finite second moments, and that*

$$\sum_{n=1}^{\infty} v_n \textit{ is convergent,} \tag{1}$$

where $v_n = \operatorname{var} X_n = E(X_n^2)$ $(n = 1, 2, \ldots)$. Then

$$\sum_{n=1}^{\infty} X_n \textit{ is a.s. convergent.}$$

Proof. Let $\quad S_n = X_1 + X_2 + \ldots + X_n \qquad (n = 1, 2, \ldots)$

and

$$A = \left\{\omega : \sum_{n=1}^{\infty} X_n(\omega) \text{ is convergent}\right\}.$$

Then $\quad A = \{\omega : \text{the sequence } S_1(\omega), S_2(\omega), \ldots \text{ is convergent}\}$

$$= \bigcap_{p=1}^{\infty} \bigcup_{m=1}^{\infty} \bigcap_{n=m+1}^{\infty} \left\{|S_n - S_m| < \frac{1}{p}\right\} \quad \text{(see page 102)}$$

$$= \bigcap_{p=1}^{\infty} \bigcup_{m=1}^{\infty} \bigcap_{n=m+1}^{\infty} B_{mpn},$$

where $\quad B_{pmn} = \left\{|S_k - S_m| < \dfrac{1}{p} \quad (k = m+1, m+2, \ldots, n)\right\}.$

Therefore

$$A^c = \bigcup_{p=1}^{\infty} \bigcap_{m=1}^{\infty} \bigcup_{n=m+1}^{\infty} B_{pmn}^c = \bigcup_{p=1}^{\infty} \bigcap_{m=1}^{\infty} C_{pm}, \quad \text{say.}$$

Now $P(B^c_{pmn}) = P\left(\max\{|X_{m+1}+\ldots+X_k|: k = m+1, \ldots, n\} \geqslant \frac{1}{p}\right)$

$$\leqslant p^2 \sum_{k=m+1}^{n} v_k \quad \text{(by Lemma 2).}$$

Therefore

$$P(C_{pm}) = \lim_{n\to\infty} P(B^c_{pmn}) \quad \text{(by Theorem 2.2)}$$

$$\leqslant p^2 \sum_{k=m+1}^{\infty} v_k.$$

For each positive integer m'

$$P\left(\bigcap_{m=1}^{\infty} C_{pm}\right) \leqslant P(C_{pm'}) \leqslant p^2 \sum_{k=m'+1}^{\infty} v_k.$$

On letting m' tend to infinity, it follows from (1) that $P\left(\bigcap_{m=1}^{\infty} C_{pm}\right) = 0$, and so $P(A^c) = 0$.

Thus $P(A) = 1$, which is the desired result.

Lemma 4. *Suppose that the infinite series* $\sum_{n=1}^{\infty} a_n$ *(of real or complex numbers) is convergent. Then*

$$\frac{1}{n}\sum_{k=1}^{n} ka_k \to 0 \quad \textit{as} \quad n\to\infty.$$

Proof. Let $s_n = a_1+a_2+\ldots+a_n$ $(n = 1, 2, \ldots)$, and

$$s = \lim_{n\to\infty} s_n.$$

Then

$$\frac{1}{n}\sum_{k=1}^{n} ka_k = \frac{1}{n}\{s_1+2(s_2-s_1)+\ldots+n(s_n-s_{n-1})\}$$

$$= \left(1+\frac{1}{n}\right)s_n - \frac{1}{n}(s_1+s_2+\ldots+s_n).$$

Since $\frac{1}{n}(s_1+s_2+\ldots+s_n) \to s$ as $n\to\infty$ (by a standard theorem on limits), the result follows.

Theorem 1 (*the strong law of large numbers*) *Suppose that* $X_1, X_2, \ldots$ *are independent observations of a random variable* X *with finite expected value* μ.

Then

$$\bar{X}_n = \frac{1}{n}(X_1+X_2+ \ldots +X_n) \to \mu \text{ a.s.} \quad \text{as} \quad n \to \infty.$$

Proof. For $n = 1, 2, \ldots$ let

$$Y_n = \begin{cases} X_n & \text{if} \quad |X_n| < n \\ 0 & \text{if} \quad |X_n| \geqslant n. \end{cases}$$

Then $P(X_n \neq Y_n) = P(|X_n| \geqslant n)$

$= P(|X| \geqslant n)$ (because X_n and X have the same distribution).

$$\text{Therefore} \quad \sum_{n=1}^{\infty} P(X_n \neq Y_n) = \sum_{n=1}^{\infty} P(|X| \geqslant n)$$

$$\leqslant E(|X|) \quad \text{(by Exercise 5.18)}$$

$$< \infty.$$

Therefore, by Lemma 1,

$$\frac{1}{n}\sum_{k=1}^{n}(X_k-Y_k) \to 0 \text{ a.s.} \quad \text{as} \quad n \to \infty. \tag{2}$$

Now, for $n = 1, 2, \ldots,$

$$E(Y_n^2) = E(X_n^2 I_{\{|X_n| < n\}})$$
$$= E(X^2 I_{\{|X| < n\}}),$$

again because X_n and X have the same distribution (see Exercise 4.5). Let

$$A_n = \{n-1 \leqslant |X| < n\}$$

and

$$B_n = A_1 \cup A_2 \cup \ldots \cup A_n = \{|X| < n\} \quad (n = 1, 2, \ldots).$$

Then
$$\sum_{n=1}^{\infty} n^{-2} \operatorname{var} Y_n \leqslant \sum_{n=1}^{\infty} n^{-2}E(Y_n^2)$$

$$= \sum_{n=1}^{\infty} n^{-2}E(X^2 I_{B_n})$$

$$= \sum_{n=1}^{\infty} n^{-2} \sum_{k=1}^{n} E(X^2 I_{A_k})$$

$$= \sum_{k=1}^{\infty} E(X^2 I_{A_k}) \sum_{n=k}^{\infty} n^{-2}$$

$$\leqslant \sum_{k=1}^{\infty} E(X^2 I_{A_k}) \times \frac{2}{k}$$

$$\leqslant 2 \sum_{k=1}^{\infty} E(|X| I_{A_k})$$

$$\text{(because } |X| < k \text{ on } A_k \; (k = 1, 2, \ldots))$$

$$= 2E\left(\sum_{k=1}^{\infty} |X| I_{A_k}\right) \quad \text{(by Exercise 5.16)}$$

$$= 2E(|X|) \quad \left(\text{because } \sum_{k=1}^{\infty} I_{A_k} = I_{\Omega} = 1\right)$$

$$< \infty.$$

Now the random variables

$$\frac{1}{n}\{Y_n - E(Y_n)\} \qquad (n = 1, 2, \ldots)$$

are independent (by Exercise 6.7 (i)). Therefore, by Lemma 3,

$$\sum_{n=1}^{\infty} \frac{1}{n}\{Y_n - E(Y_n)\} \quad \text{is a.s. convergent,}$$

and so, by Lemma 4,

$$\frac{1}{n}\sum_{k=1}^{n}\{Y_k - E(Y_k)\} \to 0 \text{ a.s.} \quad \text{as} \quad n \to \infty. \tag{3}$$

Finally $E(Y_k) = E(X_k I_{\{|X_k| < k\}}) = E(XI_{B_k})$

$$\to E(X) = \mu \quad \text{as} \quad k \to \infty$$

(by the dominated convergence theorem), and so, by a standard theorem on limits

$$\frac{1}{n}\sum_{k=1}^{n} E(Y_k) \to \mu \quad \text{as} \quad n \to \infty. \tag{4}$$

The result follows from (2), (3) and (4).

Note. Suppose that $X_1, X_2, \ldots$ are independent observations of a random variable X. Then the following extensions of the strong law may be shown to hold:

(i) If $E(X)$ exists (but is not necessarily finite), then

$$\bar{X}_n \to E(X) \text{ a.s.} \quad \text{as} \quad n \to \infty.$$

(ii) If $E(|X|) = \infty$, then

$$\overline{\lim_{n\to\infty}} \, |\bar{X}_n| = \infty \text{ a.s.}$$

Further reading on probability

A prerequisite is some knowledge of the theory of measure and integration in general, and of Lebesgue theory in particular. The former may be obtained from any one of the first three books in the following list and the latter from the foorth.

P. R. Halmos, *Measure Theory*, Van Nostrand, 1950 (Chapters 1–8)

J. F. C. Kingman and **S. J. Taylor,** *Introduction to Measure and Probability*, Cambridge University Press, 1966 (Chapters 1–9)

M. Loève, *Probability Theory* (3rd edn), Van Nostrand, 1963 (Chapters 1–2)

H. Kestelman, *Modern Theories of Integration*, Dover, 1960.

For those who wish to continue with probability theory as quickly as possible, I recommend Loève, who gives an attractive treatment of the necessary minimum.

The following four works presuppose some knowledge of measure and integration theory. Thus although each one overlaps with the present work it can, in its own way, take the subject further.

K. L. Chung, *A Course in Probability Theory*, Harcourt, Brace & World, 1968.

H. Cramér, *Random Variables and Probability Distributions* (3rd edn), Cambridge University Press, 1970.

J. Lamperti, *Probability*, Benjamin, 1966.

H. G. Tucker, *A Graduate Course in Probability*, Academic Press, 1967.

Of these I particularly recommend the very readable book by Lamperti. It also has a useful bibliography of selected references to the original literature, to serve as a guide to yet further reading.

The encyclopaedic work by Loève will be found useful for referen

A. N. Kolmogorov, *Foundations of the Theory of Probability* (2nd En edn), Chelsea, 1956.

This short work, which first appeared (in German) in 1933, has inevitably been superseded. But every probabilist should read it at least once, to see how it all began.

Last, but by no means least, the epoch-making

W. Feller, *An Introduction to Probability Theory and its Applications* (2 vols), Wiley (vol. I, 3rd edn, 1968; vol. II, 2nd edn, 1971).

Volume I is limited to discrete random variables; the general theory is treated in the mathematically more demanding volume II.

It is impossible in a brief note to do justice to the wealth of applications and mathematical ingenuity contained in these two books. Exasperating as they can be on occasion, they are nevertheless essential reading.